Essentials of Radiation Heat Transfer

C. Balaji

Essentials of Radiation Heat Transfer

Ane Books
Pvt. Ltd.

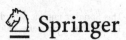

C. Balaji
Department of Mechanical Engineering
Indian Institute of Technology Madras
Chennai, Tamil Nadu, India

ISBN 978-3-030-62619-8   ISBN 978-3-030-62617-4   (eBook)
https://doi.org/10.1007/978-3-030-62617-4

Jointly published with ANE Books India
In addition to this printed edition, there is a local printed edition of this work available via Ane Books in
South Asia (India, Pakistan, Sri Lanka, Bangladesh, Nepal and Bhutan) and Africa (all countries in the
African subcontinent).
ISBN of the Co-Publisher's edition: 9789382127864

This Springer imprint is published by the registered company Springer Nature Switzerland AG
The registered company address is: Gewerbestrasse 11, 6330 Cham, Switzerland

But even if the radiation formula should prove to be absolutely accurate it would after all be only an interpolation formula found by happy guesswork, and would thus leave one rather unsatisfied. I was, therefore, from the day of its origination, occupied with the task of giving it a real physical meaning.

—Max Planck
(1919 Nobel Prize address, 'The Origin and Development of the Quantum Theory'.)

This book is dedicated to
Max Planck

Preface

This book is an outgrowth of my lectures for the courses "Conduction and Radiation" and "Radiative heat transfer" that I have been offering almost continuously since 1998 at IIT Madras. The question uppermost in the minds of many readers maybe "Why another book on radiation"? My response to this is that in every subject or course, there is still space for a new book so long as the latter is able to bring in some alacritic freshness either in the content, or the treatment or both. Through this book, I have endeavoured to "decomplexify" or more acceptably, "demystify" radiation heat transfer, which is anathema to many students. I use an easy to follow conversational style, backed up by fully worked out examples in all the chapters to vaporise the myth that radiation is only for daredevils.

Though I research quite a lot in radiation, I have scrupulously avoided adding material based on the findings of my research. The focus instead is to present a book that can be used either at the senior undergraduate or at the graduate level. Based on my past experience, I believe that the material presented in this book can be covered in about 40 lectures, each of 50 min duration. Carefully chosen exercise problems supplement the text and equip students to face "radiation" boldly.

I thank Prof. S. P. Venkatesan, my former research advisor and colleague, for introducing me to radiation and for graciously passing on to me the baton of carrying radiation forward after I joined IIT Madras.

Thanks are due to my wife Bharathi for painstakingly transcribing my video lectures on radiation offered under National Program on Technology Enhanced Learning (NPTEL), India for the course "Conduction and Radiation". This served as the starting point for the book. I thank my students Ramanujam, Gnanasekaran, Pradeep Kamath, Chandrasekar, Konda Reddy, Rajesh Baby, Samarjeet, Krishna and Srikanth for their help with the exercises and examples and in compiling the material in TEX. Special thanks are due to my doctoral student Samarjeet, who spent long hours with me in reworking the examples and the text for Chap. 7, for this international edition that was first co-published by Ane and John Wiley and now by Ane and Springer.

I also wish to thank the Center for Continuing Education, IIT Madras, for financial assistance.

The support of ANE books for bringing out the book in record time is gratefully acknowledged. I also thank Springer for coming forward to take this book to the international markets.

This edition is an updated version of the book where small typos and bugs here and there have been fixed. A few minor edits have also been done, wherever they were required to enhance clarity. I would like to thank Sandeep, Sangamesh, Pradeep, Shashikant, Girish and Rahul for proofing the Springer edition at express speed.

Thanks are also due to my daughter Jwalika for being so understanding.

I would be glad to respond to queries and suggestions at balaji@iitm.ac.in.

Chennai, India   C. Balaji
December 2020

Contents

About the Author

Dr. C. Balaji is currently the T.T. Narendran Chair Professor in the Department of Mechanical Engineering at the Indian Institute of Technology (IIT) Madras, India. He graduated in Mechanical Engineering from Guindy Engineering College, Chennai, India in 1990 and obtained his M.Tech (1992) and Ph.D. (1995) from Indian Institute of Technology Madras, India in the area of heat transfer. His areas of interest include heat transfer, computational radiation, optimization, inverse problems, satellite meteorology and atmospheric sciences. He has more than 200 international journal publications to his credit and has guided 30 students so far. Prof. Balaji has several awards to his credit and notable among them include Young Faculty Recognition Award of IIT Madras (2007) for excellence in teaching and research, K. N. Seetharamu Award and Medal for excellence in Heat Transfer Research (2008), Swarnajayanthi Fellowship Award of the Government of India (2008–2013), Tamil Nadu Scientist Award (2010) of the Government of Tamilnadu, Marti Gurunath Award for excellence in teaching (2013) and Mid-Career Research Award (2015) both awarded by IIT Madras. He is a Humboldt Fellow and an elected Fellow of the Indian National Academy of Engineering. Prof.Balaji has authored 8 books thus far. He is currently the Editor-in-Chief of the International Journal of Thermal Sciences.

Chapter 1
Introduction

1.1 Importance of Thermal Radiation

Heat transfer is accomplished by one or more of the following modes namely, conduction, convection and radiation. However, the basic modes of heat transfer are only
two: conduction and radiation, as convection is a special case of conduction where
there is a macroscopic movement of molecules outside of an imposed temperature
gradient. We restrict our attention to radiation heat transfer in this book.

Now we look at the importance of thermal radiation. Most people have the feeling
that thermal radiation is important only if the temperatures are high. Generally, when
temperatures are low, "radiation can be neglected" is the familiar refrain or argument
put forward by many people who are not inclined to include it in their analysis.
We will consider an example very shortly and try to find out if this assumption of
neglecting radiation in heat transfer analysis is justified or not.

Exploring the relation between the heat transfer rate and the temperature gradient,
we have

$$Q_{\text{cond}} \propto \Delta T \tag{1.1}$$

$$Q_{\text{conv}} \propto \Delta T \tag{1.2}$$

Equation (1.2) is strictly not valid for free convection. Let us consider, a frequently
used correlation for the dimensionless heat transfer coefficient, namely the Nusselt
number, for free convection.

$$Nu = aRa^b \tag{1.3}$$

$$Nu = a(c\Delta T)^b \tag{1.4}$$

$$b \approx 0.25 \text{ (for laminar flow)} \tag{1.5}$$

$$Nu \propto (\Delta T)^{0.25} \tag{1.6}$$

$$q_{\text{conv}} \propto (\Delta T)^{1.25} \tag{1.7}$$

© The Author(s), under exclusive license to Springer Nature Switzerland AG 2021
C. Balaji, *Essentials of Radiation Heat Transfer*,
https://doi.org/10.1007/978-3-030-62617-4_1

So q is proportional to ΔT to the power of 1.25 for laminar natural convection flows. For turbulent natural convection flows, q will go as $(\Delta T)^{1.33}$.

The Rayleigh number, Ra in Eq. (1.3), is given by

$$Ra = \frac{g\beta\Delta T L^3}{\nu\alpha} \tag{1.8}$$

where

- g—acceleration due to gravity
- β—isobaric cubic expansivity (for ideal gases, β can be equated to 1/T, where T is the temperature in Kelvin)
- ΔT—temperature difference imposed in the problem
- L—characteristic dimension, which can be the length of a plate or the diameter of a cylinder or sphere
- ν—kinematic viscosity
- α—thermal diffusivity.

Now we can see that q_{conv} is proportional to ΔT with a pinch of salt as it is actually $(\Delta T)^{1.25}$ for natural convection. For radiation

$$q_{rad} = (T^4 - T_\infty^4) \tag{1.9}$$

A non linearity enters the problem right away because q is proportional to the difference in the fourth powers of temperature.

Assume that a bucket filled with water is heated using an immersion heater. Then under steady state, we know that Q_h heat supplied must equal that lost to the outside as the sum of the convective and radiative heat transfer, then

$$Q_h = Q_{conv} + Q_{rad} \tag{1.10}$$
$$Q_h = hA(T - T_\infty) + \epsilon\sigma A(T^4 - T_\infty^4) \tag{1.11}$$

In Eq. 1.11, ϵ is the emissivity of the surface and A is the surface area (we will study about emissivity in far greater detail in a later chapter). Equation 1.11 has to be solved iteratively even under steady state to determine the temperature of the water with the bucket, assuming that the material of the bucket is at the same temperature as that of the water. To solve this non- linear equation, we need to assume a value of T of water and see if the LHS is equal to the RHS. If they are not equal, then we update the value of T and redo the procedure and this is repeated till the LHS becomes equal to the RHS. This is called the *successive substitution method*.

The difficulty with radiation first stems from the fact that radiation is proportional to $(T^4 - T_\infty^4)$. Therefore, *its importance increases non linearly with increasing temperature*. So at high temperatures of the order of 1200 °C or 1500 °C, whether it is an IC engine, furnace or boiler, there is no escape from considering radiation, as this will be the dominant mode of heat transfer. In fact, in boilers, there is a radiant superheater section, where the ultimate heat transfer takes place and the temperature of the steam is lifted. Even in the ubiquitous microwave oven, there is basically radiative heating in the microwave region of the spectrum. The importance of thermal radiation first stems from the fact that q_{rad} varies non linearly with temperature.

The second point is that *radiation requires no material medium to propagate*. The proof is the receipt of solar radiation on this earth from the sun, which lies millions of miles away. This shows that radiation is able to travel through vacuum. In fact, radiation travels best in vacuum, because there is no absorption or scattering. Once it enters the earth's atmosphere, there is absorption and reflection by certain molecules. This reflection is called scattering. Also, as these molecules are at a temperature greater than 0 K, as a consequence of the Prevost's law, they also emit. So the atmosphere is emitting, absorbing and scattering. However, outside the atmosphere, the radiation is able to travel without any distortion at all.

The third point is that even at low temperatures, radiation may be significant. Let us consider an example. We consider a vertical flat plate whose length, L = 0.5 m and is maintained at T_w = 373 K standing in still and quiescent air with emissivity 0.9 (i.e. it is coated with black paint). The ambient temperature, T_∞ = 303 K. Needless to say, a natural convection boundary will be set up along the plate, on both sides. The boundary layer will develop as shown in Fig. 1.1. The velocity at points A and B will be 0 for different reasons. At A, the velocity is zero as a consequence of the no slip condition, while at B, it is zero, because air is quiescent in the free stream region.

The Nusselt number is given by

$$Nu = aRa^b \qquad (1.12)$$

Let us consider a very simple, well known correlation for laminar natural convection from a vertical plate, where a and b are 0.59 and 0.25 based on well known results from, Sparrow and Gregg (1956).

$$Nu = 0.59Ra^{0.25} \qquad (1.13)$$

The Rayleigh number is calculated with the following values: $g = 9.81$ m/s^2, ΔT = 70 K, $\beta = 1/T_{mean}$, $T_{mean} = (373+303)/2 = 338$ K, $\nu = 16 \times 10^{-6}$ m^2/s ; Pr = $\nu/\alpha = 0.71$. The Rayleigh number, Ra turns out to be 7×10^8. (When $Ra < 10^9$, the flow is laminar and when Ra$> 10^9$, transition to turbulent flow begins.) Substituting for Ra in Eq. 1.13, Nu = 96. The Nusselt number is the dimensionless heat transfer coefficient, which is given by Nu = hL/k; k = 0.03 W/mK for air. We now calculate the average heat transfer coefficient of the plate, h to be 5.8 W/m^{2}K.

Although both radiation and convection are taking place on both sides of the plate, let us consider for the present that they take place from just one side. (q_{conv} = hΔT = 5.8×70 = 406 W/m^2). Such calculations are also very profound as the temperature we are talking about, 100 °C, is more than the reliable temperature of operating electronic equipment, which is normally about 80 or 85 °C. So if we do all these calculations, we get q_{conv} = 406 W/m^2.

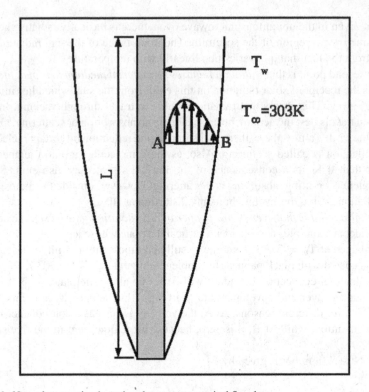

Fig. 1.1 Natural convection boundary layer over a vertical flat plate

At this point, a little digression is in order. So let us now say we have some other situation where we have h to be 6 or 6.5. q_{conv} can touch about 500 W/m^2 in this case. So we are talking about flux levels of 0.5 kW/m^2 of natural convection. If we are talking about a flux level in our equipment which is more than 0.5 kW/m^2, we have to use a fan to cool it in order to maintain it at the desired temperature!

We can also do similar calculations and determine the maximum flux that one fan can withstand. If required we use 2 fans, similar to what is found in desktop computers. After that comes liquid cooling, impingement cooling. For example, data centres cannot be cooled by fans alone. The air itself will be conditioned such that the data centre is maintained at, say 16 °C.

Getting back to the problem at hand, we need to find out what the q_{radn} for this problem will be. The following assumptions hold, (1) the sink for the radiation is the same as the sink for convection, (2) the walls of the room are at the same temperature as the ambient, which is a reasonable assumption. (Sometimes, T_∞ for convection need not be the same as T_∞ for radiation. But most of the times, we assume them to be the same.), (3) Stefan Boltzmann constant $\sigma = 5.67 \times 10^{-8}$ W/m^{2}K^4; $\epsilon = 0.9$. For these values, $q_{rad} = 557$ W/m^2.

$$\therefore q_{\text{total}} = q_{\text{conv}} + q_{\text{rad}}$$

$$= 964 \text{ W/m}^2 \tag{1.14}$$

$\therefore$ The radiation contribution is

$$\frac{q_{\text{rad}}}{q_{\text{tot}}} = \frac{567}{964}$$

$$= 57.8 \tag{1.15}$$

This clearly proves that radiation cannot be neglected at low temperatures.

This analysis has, however, be taken with a pinch of salt. Suppose we blow air using a fan, wherein the natural convection will change to forced convection, the heat transfer coefficient instead of being 5, may change to a value of 15 or 20. Then q_{conv} may have a value of 1 or 1.2 kW/m^2. So convection will begin dominating radiation. However, even if the flux level is 1.8 kW/m^2, $q_{\text{rad}}/q_{\text{total}}$ is not negligibly small. So radiation may be neglected only in cases where the other modes of heat transfer are dominant. If it is convection in air, free or forced, radiation cannot be neglected. Even so, if the medium is water, the story changes completely. Water has a terrific thermal conductivity of 0.6 W/mK as opposed to air. All these numbers will change because the Nusselt number is hL/k. Since h increases for water, the radiation contribution will be negligible. The radiation treatment also gets messy as water "participates" in the radiation. So if we have air cooling and are doing computational fluid dynamics (CFD) analysis of a desktop or some other electronic equipment, we cannot neglect radiation in our analysis. Thankfully, commercial software has radiation modules and many people use the combined analysis nowadays in the prediction of maximum or operating temperatures of electronic equipment. In summary, in natural convection alone or in mixed convection, where both natural and forced convection are important, radiation plays a part and cannot be neglected by simply putting forward the argument that temperature is very low.

1.2 Nature of Radiation

To explain radiation and its effects, generally, two models are used (i) the wave model and (ii) the quantum model. Using the wave model, we can characterize radiation by wavelength, frequency and speed; all that which is applicable for optics can be applied here too but neither the radiative properties of gases nor black body behaviour could be explained using the electromagnetic theory, and hence the quantum theory had to be developed. Electromagnetic radiation travels at the speed of light. Therefore, the velocity of light in vacuum c_o can be assumed to be the velocity of electromagnetic radiation in vacuum. $c_o = 2.998 \times 10^8$ m/s or 3×10^8 m/s (app.).

Now we can characterize radiation by the following additional parameters: ν— frequency, λ—wavelength, $1/\lambda$—wave number. If the velocity of light in a medium

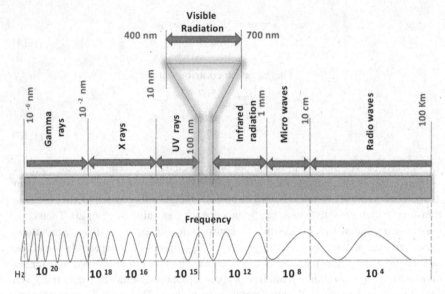

Fig. 1.2 Electromagnetic spectrum

is c, we know that c must be less than or equal to c_o. The refractive index of the medium $= n = c_o /c$. For glass, n $= 1.5$ and for gases, n ≈ 1;

Now let us look at the electromagnetic spectrum which can be characterized by either (1) the wavelength, (2) the frequency (see Fig. 1.2). For example, the wavelength of radio waves is about 10^3 m. The wavelength of gamma rays is about 10^{-12} m, which gives them a high frequency of around 10^{20} Hz. The energy of electromagnetic radiation is given by $E = h\nu$ (which we shall derive later), where h, Planck's constant $= 6.626 \times 10^{-34}$ Js.

If we consider gamma rays, their energy is very high. Looking at the other end of the spectrum where we encounter radio waves, the energy is very low. This is used by electronics and communications engineers where the original signal, having low energy, is first modulated with a high energy carrier wave, transmitted and demodulated at the other end. Mechanical engineers lie somewhere between these two ends and operate in the visible, ultraviolet or infrared regions because this corresponds to reasonable levels of temperatures encountered in engineering applications. We usually are not concerned with temperatures of 10^5 or 10^6 K. The only place where we may come across this is in nuclear fusion. Generally, we talk about temperatures in the range of 200–3000 K. So, the wavelength of thermal radiation of interest to thermal engineers is $\lambda = 0.1$–100 μm.

In the visible range, whose wavelength lies between 0.4 and 0.7 μm, colours range from violet to red. For us, mechanical engineers, wavelengths of the order of 10 m are very big. We work with rays whose wavelengths range from micrometers to nanometers.

Example 1.1 Radiation at a wavelength, $\lambda = 3$ µm travels through vacuum into a medium with refractive index, n = 1.4.

(a) Determine the speed, frequency and wave number for radiation in vacuum.
(b) Determine the above quantities and also the wavelength for radiation in the medium..

Solution:

a. In vacuum:

$$c_o = 2.998 \times 10^8 \text{ m/s} \tag{1.16}$$

$$\lambda = 3 \times 10^{-6} \text{ m} \tag{1.17}$$

$$v = \frac{2.998 \times 10^8}{3 \times 10^{-6}} = 9.993 \times 10^{13} \text{ Hz} \tag{1.18}$$

$$\text{Wave number} = \frac{1}{\lambda} = \frac{1}{3 \times 10^{-6}} = 3.33 \times 10^5 \text{ m}^{-1} \tag{1.19}$$

b. In the medium:
   Even when the radiation moves from vacuum to the medium, the frequency does not change, only the wavelength changes.

$$\text{Frequency} = v = 9.993 \times 10^{13} \text{ Hz (remains same)} \tag{1.20}$$

$$c = \frac{c_0}{n} = \frac{2.998 \times 10^8}{1.4} = 2.14 \times 10^8 \text{ m/s} \tag{1.21}$$

$$c = v\lambda \tag{1.22}$$

$$\text{Wave length, } \lambda = \frac{c}{v} = 2.14 \text{ µm} \tag{1.23}$$

$$\text{Wave number} = 4.66 \times 10^5 \text{ m}^{-1} \tag{1.24}$$

Example 1.2 The wavelength and speed of radiation travelling within a medium are 3.2 µm and 2.3×10^8 m/s, respectively. Determine the wavelength of the radiation in vacuum.

Solution:

$$c = v\lambda \tag{1.25}$$

$$\text{so, } v = \frac{c}{\lambda} = \frac{2.3 \times 10^8}{3.2 \times 10^{-6}} = 7.18 \times 10^{13} \text{ Hz} \tag{1.26}$$

This is the frequency of the radiation in the medium, as well as in vacuum. The wavelength in vacuum is

$$\lambda = \frac{c_0}{v} = \frac{3 \times 10^8}{7.18 \times 10^{13}} = 4.18 \text{ µm} \tag{1.27}$$

Chapter 2
Black Body and Its Characteristics

We will now look at a very important concept in radiation heat transfer, namely, the black body.

Definition 2.1 A black body is one that allows all incident radiation and internally absorbs all of it.

So what does it imply technically? Reflection = 0. Transmittance = 0. This definition requires further qualification because that a black body allows all incident radiation and absorbs all of it is true for (a) all wavelengths (b) all incident directions. Therefore, for a given wavelength and in a given incident direction, there can be no body which absorbs more radiation than a black body. Any real body has to absorb radiation which is lower than that of the black body. This is essentially a conceptual definition and serves as a benchmark. This concept of maximum absorption, regardless of the incident wavelength and direction is central to the understanding of radiative heat transfer. So, a black body serves as the benchmark or the gold standard against which all other real surfaces can be compared.

Incidentally, the black body also emits the maximum radiation for a given temperature. This is a consequence of the above, which can be proved. The black body is basically defined based on its ability to absorb fully and not on its emission capacities. That a black surface emits the maximum is a consequence of it being the perfect absorber and is just a corollary!

Some examples of black bodies:

- Lamp black
- Platinum black
- Gold black
- Special paints.

Suppose we want to do experiments using a black body in the laboratory, we usually want to take an aluminium plate and coat it with black paint. However, the emissivity

© The Author(s), under exclusive license to Springer Nature Switzerland AG 2021

C. Balaji, *Essentials of Radiation Heat Transfer*, .

https://doi.org/10.1007/978-3-030-62617-4_2

of this, when measured, will be just 0.8 or 0.88 or 0.9 because there will be some places where we have not fully coated it. After the second coat of paint, the emissivity may increase to 0.92 or so. After that, the emissivity will asymptotically saturate; ideally, we cannot get to 1.00. We can get a maximum of 0.94 or 0.95 and for all practical purposes, this can serve as a black body or stated clearly, can serve as a high emissivity surface.

Let us move on to an interesting question—why the name "black body"? Generally, black bodies are very poor reflectors. Hence, they "appear" to be "visually black"! Even so, the eye is a very poor instrument to detect radiation because it can detect radiation only in a very narrow range of 0.4–$0.7\mu m$. So a surface may be very black in the range 0.4–$0.7\mu m$, but in the other parts of the spectrum its "blackness" cannot be visually evaluated and hence verified, and we need sophisticated equipment like the spectrometer to determine its behaviour.

On the contrary, since the visible part of the spectrum is genuinely a part of the electromagnetic spectrum, if something is truly radiatively black, it will be black between 0.4 and 0.7 μm too. Therefore, all radiatively black bodies have to be visually black.

So a radiatively black body will be visually black, but a visually black body need not necessarily be radiatively black.

2.1   Key Attributes of a Black Body

- **Perfect emitter** (perfect absorber is already there in the definition)
  Let us consider an evacuated enclosure which is at a temperature T_∞ with vacuum inside, as shown in Fig. 2.1. Now let a small black body, initially at temperature T_w, be inserted into the middle of the enclosure. Let $T_w > T_\infty$.

Since the black body is not touching the walls of the enclosure, there is no conduction heat transfer. Since the chamber is evacuated and there is no medium, convection is also non-existent. Let us say $T_\infty = 30\,°C$, while $T_w = 200\,°C$. Since this is a small body in a large enclosure, after sufficient time has elapsed, the black body will achieve thermal equilibrium with the surroundings, i.e. it will reach a temperature of T_∞, as shown in Fig. 2.2. The small object is, however, a black body and the question to ask here is—"What is the story here?" The small object is absorbing exactly the same amount as it is emitting because if the emission is not equal to the absorption, then there is a net rate of change of enthalpy which has to take place inside the black body, as a consequence of which, its temperature has to go down or go up, which is again forbidden by the second law of thermodynamics because equilibrium has already been established. Therefore, the amount of radiation which is emitted by the black body is (or has got to be) exactly equal to the amount of radiation which is absorbed by the black body. Since the body under consideration is black and is absorbing the maximum amount of radiation, therefore, it is also emitting the maximum amount of radiation!

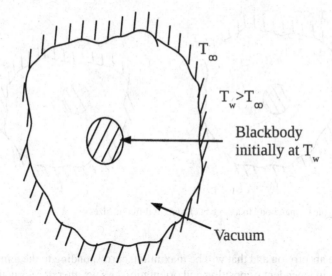

Fig. 2.1 Illustration for proving that a black body is a perfect emitter

Fig. 2.2 Temperature time
history of the small black
body undergoing cooling in a
large enclosure

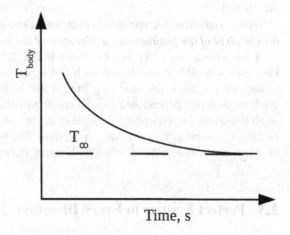

2.2 Radiation Isotropy

Consider an enclosure with a black body similar to the one shown in Fig. 2.1. Now
let this black body be placed in another enclosure of a size smaller than the previous
one. The two cases are shown in Fig. 2.3a, b, respectively. The enclosure temperature
is the same for both the cases and so is the initial temperature of the black body.

So long as the enclosure temperature is T_∞ and both are small bodies placed in
large surroundings (which have an infinite capacity to take on the heat), regardless
of the position of the black body, both will reach the same equilibrium temperature
upon cooling of the black bodies. Upon reaching equilibrium, the emission will be

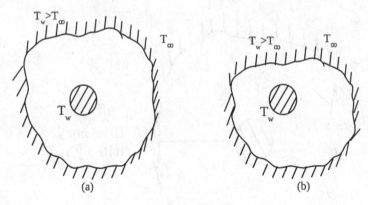

Fig. 2.3 A typical small body in a large enclosure of different sizes

equal to the absorption and that will be maximum, corresponding to the temperature T_∞. This is independent of position and orientation and we, therefore, say that radiation isotropy exists within the enclosure and that a uniform radiation field has been established.

Hence, radiation isotropy means that the radiation field within the enclosure is independent of the position and orientation of the black body.

If we make a cavity like the one shown in Fig. 2.4, and close it on all sides and have only a small hole and heat it such that it becomes an evacuated cavity that is heated, the radiation field emerging from it will be isotropic, meaning that it does not have direction dependence and the radiation comes out with uniform intensity in all directions corresponding to the temperature of the body. This is known as a *Hohlraum* meaning "empty room" in German. The hohlraum concept can be used to mimic or simulate a "near" black body under laboratory conditions.

2.3   Perfect Emitter in Every Direction

Consider an enclosure as shown in Fig. 2.5, with a small area dA being active on the walls of the enclosure with all the other areas being radiatively inactive. Even for this situation, after sufficient time has elapsed, equilibrium will be established and the body will be cooled down to the temperature T_∞.

Now, the body will continue to absorb radiation which will be the maximum as it is black. Even so, all the radiation and this radiation will be maximum as it is a black body but all the radiation is coming in a particular direction because only one portion of the enclosure wall is active. The black body has to radiate back the same radiation for equilibrium to be established. Therefore, since it is absorbing the maximum in that particular direction, it has to radiate a maximum in the same direction. Therefore, in a particular direction, it will be a maximum emitter too. Since it is anyway the same in all directions, *this emitted radiation is maximum and equal in all directions.*

Fig. 2.4 The schematic of
the cavity used for
illustration of radiation
isotropy

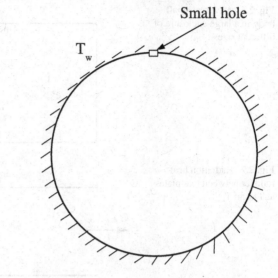

Fig. 2.5 Schematic to prove
that the black body is a
perfect emitter in all
directions

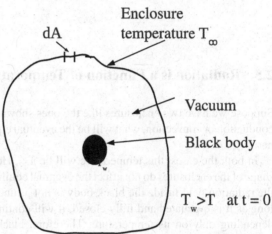

2.4 Perfect Emitter in Every Wavelength

Now we can repeat the same experiment such that the walls of the enclosure are so designed that they emit or absorb radiation in very small intervals of $d\lambda$ about λ. The black body will also absorb radiation in a small wavelength interval $d\lambda$ about λ. While it can continue emitting radiation in any other wavelength, the walls of the enclosure are in a position to absorb radiation only in the wavelength $d\lambda$ about λ. Therefore, whatever is absorbed must be equal to whatever is emitted in order that equilibrium is maintained. This $d\lambda$ about λ is purely under our control. It should be valid for any $d\lambda$ about λ. Hence, at every wavelength, the black body will be a perfect emitter.

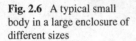

Fig. 2.6 A typical small body in a large enclosure of different sizes

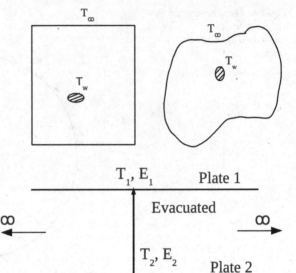

Fig. 2.7 Radiation heat transfer between two plates at T_1 and T_2

2.5 Radiation is a Function of Temperature Alone

Suppose we have two enclosures like the ones shown in Fig. 2.6, evacuated with no conduction or convection, what will be the eventual equilibrium temperature reached here?

In both the cases, this temperature will be T_∞. Hence, the characteristics or the shape of the enclosures do not affect the eventual equilibrium temperature. Therefore, the radiation field inside the black body is not a function of the shape and size. So long as it is evacuated and fully closed, it will continue to emit isotropic radiation, depending only on its temperature. Therefore, black body radiation strength is a function of T alone.

2.6 Does Radiation Strength Increase or Decrease With Temperature?

The answer is obvious that it increases with temperature. To prove it from thermodynamic arguments, consider two plates which are at temperatures T_1 and T_2 and whose radiation strengths are E_1 and E_2, as shown in Fig. 2.7. Let us now assume that $T_1 > T_2$, while $E_1 < E_2$. $Q = E_2 - E_1$ and hence the direction of flow of energy will be from plate 2 to plate 1. Positive transfer of energy from a body at a lower temperature

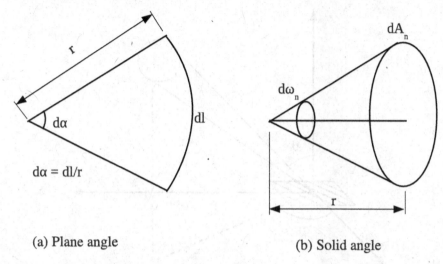

Fig. 2.8 Depiction of plane and solid angles

to a body at a higher temperature is forbidden by the law of thermodynamics unless we do some work. Hence, nothing in this argument is wrong except for the initial assumption that E is proportional to $1/T$. Therefore, the original assumption should be wrong which proves that E should be a monotonically increasing function of temperature. Such a way of proving something by proving the converse to be absurd is called *Reductio as absurdum*, and has been extensively used in the past.

These are the attributes of a black body. But what is this E ? How is it related to the temperature, we do not know. The best of physicists were working on this problem a little over hundred years ago. In order to derive the quantitative aspects of the black body behaviour, it is imperative that we study some solid geometry.

2.7 Solid Angle—dω

Consider Fig. 2.8a in which the plane angle is shown and is given by $d\alpha = dl/r$. Consider the elemental area dA_n which subtends an elemental solid angle $d\omega_n$ as shown in Fig. 2.8b. The elemental solid angle $d\omega_n$ is given by

$$d\omega_n = \frac{dA_n}{r^2} \text{ steradians or sr}$$

(2.1)

where A_n is the normal area.

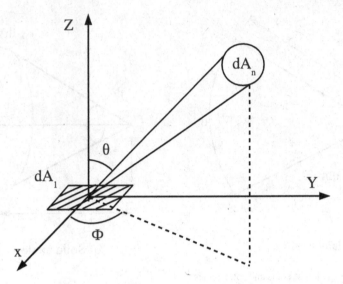

Fig. 2.9 Emission of radiation from a differential area dA_1 and intercepted by another differential area dA_n

2.7.1 Spherical Coordinate System

Let us consider an elemental area dA_1, as seen in Fig. 2.9, that is emitting radiation in all directions. There is another elemental area dA_n which is intercepting this radiation and if we shine a torchlight on dA_n, its shadow will fall on the plane such that the angle thus formed is called the **azimuthal angle** denoted by ϕ. The other angle formed will be θ, which is measured from the vertical, as seen in Fig. 2.9. This angle is known as the **zenith angle**. If we have the axes x, y and z and we have a point (x_1, y_1, z_1) as shown in Fig. 2.10, the coordinates of this point can also be described as (r, θ, ϕ) where $r^2 = x_1^2 + y_1^2 + z_1^2$. We introduce the spherical coordinate system because it is operationally convenient for us to work with this in radiation heat transfer.

2.7.2 Solid Angle Subtended by an Elemental Area dA_n in the Spherical Coordinate System

Consider Fig. 2.11. The solid angle is the angle subtended by an elemental area dA_n at a point on dA_1 where dA_1 is the area which is emitting the radiation and dA_n is the elemental area which is receiving the radiation. So, for defining the solid angle we need a giver and a taker (for radiation). dA_1 is the emitting surface and radiation is spreading from this surface in all directions. Among all the directions, we are taking a small elemental area dA_n and are trying finding out how much radiation this area

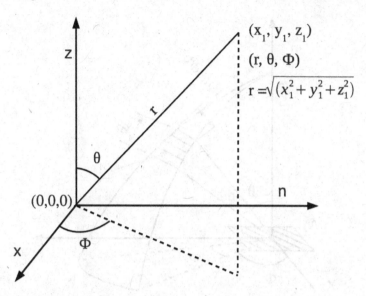

Fig. 2.10 The spherical coordinate system

intercepts and try and work this out in terms of the fundamental coordinates. As r keeps increasing, it is intuitively apparent that for the same area, the fraction of radiation captured will keep decreasing.

$$d\omega = \frac{dA_n}{r^2} \tag{2.2}$$

$$dA_n = r \sin\theta \, d\phi \, r \, d\theta \tag{2.3}$$

$$d\omega = \frac{r^2 \sin\theta \, d\phi \, d\theta}{r^2} \tag{2.4}$$

Therefore, the total solid angle associated with the elemental area dA_1, if radiation is falling on a hypothetical hemisphere above it, is given by

$$\omega = \int d\omega = \int_0^{2\pi} \int_0^{\frac{\pi}{2}} \sin\theta \, d\theta \, d\phi \tag{2.5}$$

$$\omega = 2\pi [-\cos\theta]_0^{\pi/2} \tag{2.6}$$

$$\omega = 2\pi \ \text{sr} \tag{2.7}$$

Therefore, the total solid angle associated with the hemisphere is 2π sr. The solid angle associated with a sphere will be 4π sr.

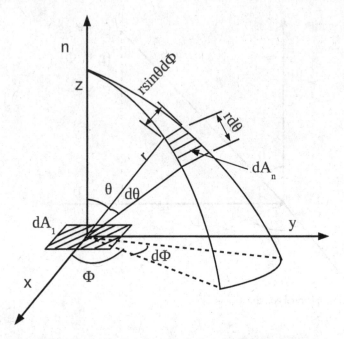

Fig. 2.11 Solid angle subtended by dA_n about dA_1 in the spherical coordinate system

The solid angle is a very important concept in radiative heat transfer because we are finally interested in the radiation heat transfer between surfaces, be it the combustion chamber of an internal combustion engine or the radiant superheater of a boiler. We see that there are various surfaces, some hot and some cold. Radiation comes out of the hot surface, while water tubes are present on the cold surface. Hot water or steam flows through the tubes and the job of the radiant superheater is to heat up the steam or the water which is on one side by the radiation from the other side. So, in all these cases, we are looking at radiation heat transfer between finite surfaces. We do not always encounter a situation of a small body surrounded by a hemispherical bucket or basket. Therefore, if we are interested in heat transfer between surfaces and these surfaces are of finite area, because radiation has a tendency to spread in all the directions, it is important for us to know the directional orientation of one surface with respect to the other surface. Or in other words, we want to know how the receiving surface is oriented directionally with regard to the emitting surface. In order to do this, not only is the spherical coordinate system useful, the definition of the solid angle also becomes imperative and essential.

We already saw that ω for a hemisphere $= 2\pi$ sr, but why are we talking mostly about hemispheres and not spheres? The answer is we are interested in the radiation from a surface (at least for now!). But this is not the be all and end all of everything. For example, if we were to consider radiation from the atmosphere, then radiation will travel in the upper and lower hemispheres also. So, once we encounter radiation in participating media, as for example, in the atmosphere or in the gases of a combustion chamber, we need to consider the full sphere.

First, we want to be able to calculate the radiation and heat transfer between surfaces. There are surfaces at different temperatures, characterized by different surface properties and are oriented in different directions. The key engineering question is What is the net heat transfer between any 2 surfaces? A more involved version of this could be that, outside of radiation, conduction and convection are also present in the problem. In such an eventuality, the problem becomes a multimode heat transfer problem, which is often the case, and our goal is to be able to compute the total heat transfer.

2.8 Spectral or Monochromatic Radiation Intensity, $I_{\lambda,e}$

Figure 2.11 is crucial for understanding the concept of radiation intensity $I_{\lambda,e}$. Here dA_1 is the emitting surface or the surface that emits the radiation and the shaded region represents the elemental area dA_n that intercepts the radiation from dA_1. The centres of dA_1 and dA_n are joined, and the distance between them is denoted by radius r. In principle, if we join any point on dA_1 with any point on dA_n, that should also be r because dA_1 and dA_n are infinitesimally small surfaces. The zenith angle and the azimuthal angle are also marked on the figure. The zenith angle θ varies from 0 to $\pi/2$ for the hemisphere while the azimuthal angle ϕ varies from 0 to 2π.

Now, with this in the background, we will introduce a quantity called spectral radiation intensity denoted by $I_{\lambda,e}$ where "λ" denotes that it is a spectral quantity or that it concerns a radiation intensity in a small wavelength interval $d\lambda$ about λ. The subscript "e" denotes that emission is under consideration. In conduction and convection heat transfer, we deal with the quantity known as flux denoted by q, whose units are W/m^2. This logically leads to the question, when we already have this quantity in W/m^2, where is the need to introduce another quantity called intensity? Why cannot we work with flux and why did people deem it fit or necessary to introduce this quantity I? The answer to this question lies in the fact that radiation falling on a surface can come from all possible directions. The radiation, whether emission or reflection, in general, will be a function of wavelengths. The radiation emitted by a surface can also be in all possible directions or wavelength. Therefore, it is important that the directional and spectral nature of the intensity of radiation be taken into account and since it is very difficult to work with flux, we introduce a "radiation intensity" which takes care of spectral and directional effects. The latter, in conjunction with the solid angle and the spherical coordinate system, gives us an eminently convenient platform to begin our study of black body behaviour and radiation heat transfer itself. The spectral radiation intensity $I_{\lambda,e}(\lambda, \theta, \phi)$ is given by

$$I_{\lambda,e}(\lambda, \theta, \phi) = \frac{dQ}{dA_1 \cos\theta d\omega d\lambda} \tag{2.8}$$

The units of $I_{\lambda,e}$ will be W/m^2 μm sr.

The formal definition of spectral radiation intensity of emission, $\mathbf{I}_{\lambda,e}$ **is the rate at which radiant energy is emitted by a surface, per unit area normal to the surface, in the direction** θ, **per unit solid angle dω about** (θ, ϕ) **in the unit wavelength interval dλ about** λ.

The importance of Eq. (2.8) is as follows. If we know the distribution of $I_{\lambda,e}$, it is possible for us to integrate and determine the value of q. Furthermore, it is instructive to mention here that q is based on the actual area as it has the unit W/m^2. However, I is based on the projected area. The above equation is valid for emission, reflection or incoming radiation (also called irradiation). Therefore, we can say that this equation is a generic expression to convert I to q and it is applicable for radiation that is emitted from a body, radiation that is incident on a body and the radiation that is reflected from a body.

2.9 Spectral Hemispherical Emissive Power

Let us now get back to the black body and see how we can define its emissive power based on the framework proposed here. Then we will come to a state where if we know what $I_{\lambda,e}$ is for a black body, we can calculate the spectral flux, the directional flux, the total flux and so on. The search for the correct $I_{\lambda,e}$ produced many Nobel laureates and many celebrated physicists miserably failed to get it right because they tried to derive it from using classical physics. The defining moment arrived when Planck proposed the quantum hypothesis in order to derive the correct distribution for $I_{\lambda,e}$. Even today, we cannot say that this is the only distribution which is correct. It may be disproved later on. But the argument is that the only distribution that agrees with experiments is the Planck's distribution, and therefore, it must be correct, till it is found to be incorrect More on this later! Let us now get back to our goal of getting an integrated quantity from the definition of spectral intensity.

The spectral emissive power from a black body $E_b(\lambda)$ is

$$E_b(\lambda) = \int\limits_0^{2\pi} \int\limits_0^{\pi/2} I_{\lambda,e} \cos\theta \sin\theta d\theta d\phi \qquad (2.9)$$

This is called the spectral hemispherical emissive power. It is spectral because it is still a function of λ as we have not integrated it with respect to λ and it is hemispherical because we have integrated it with respect to a hypothetical hemisphere by doing two integrations, one with respect to θ and the other with respect to ϕ. The units of $E_b(\lambda)$ will be W/m$^2\mu$m. Since the black body is a diffuse emitter, $I_{\lambda,e}$ is not a function of θ and ϕ. Hence, one can pull out the integral and integrate the remaining expression in Eq. (2.9). Now, the beauty is that after defining the solid angle and after having introduced the spherical coordinate system and $I_{\lambda,e}$, we have an excellent framework with which we can calculate the flux. All the quantities of engineering interest are on the left hand side of the equations, be it $E_b(\lambda)$ or E_b. The right side contains I. Hence, one can say that the right hand side of Eq. (2.9) is physics while the left side is engineering!.

The hemispherical total emissive power of the black body will be a function of temperature T (please refer to our arguments in Sect. 2.6)

$$E_b(T) = \int\limits_{0}^{\infty} E_b(\lambda)d\lambda \tag{2.10}$$

Substituting for $E_b(\lambda)$ from Eq. 2.9

$$E_b(T) = \int\limits_{0}^{\infty} \int\limits_{0}^{2\pi} \int\limits_{0}^{\pi/2} I_{b,\lambda}(\lambda, T) \cos\theta \sin\theta d\theta d\phi d\lambda \tag{2.11}$$

As a black body is diffuse, $I_{b,\lambda}(\lambda,T)$ can be pulled out of the integral and retained while performing the integration with respect to the direction.

$$E_b(T) = I_b(T) \int\limits_{0}^{2\pi} \int\limits_{0}^{\pi/2} \cos\theta \sin\theta d\theta d\phi \tag{2.12}$$

$$\boxed{E_b(T) = \pi I_b(T)} \tag{2.13}$$

In Eq. (2.11), E_b represents the total hemispherical emissive power where total means that the integration is with respect to the wavelength, hemispherical means that the integration is with respect to the angle. Hence, the calculation of the total hemispherical emissive power involves three integrations. Having studied the Stefan–Boltzmann's law (from the first course in heat transfer), we must remember that in this law, 3 integrations have already been done in: θ, ϕ and λ. The primordial relation on which the Stefan-Boltzmann law is based is the fundamental quantity $E_{b,\lambda}(\lambda)$ which the black body is supposed to emit and which can be verified by experiments and has been proposed by the theory. Two points have to be reiterated here :

1. If we look at radiation transfer between surfaces, we are talking about the hemisphere and not about the sphere.
2. Surprisingly, the result in Eq. (2.13) has π and not 2π.

Example 2.1 A surface of area $A_1 = 2 \times 10^{-4}$ m^2 emits diffusely (same in all the directions). The total hemispherical emissive power from this surface is 9×10^4 W/m^2. Another small surface $A_2 = 7 \times 10^{-4}$ m^2, is oriented as shown in the Fig. 2.12. Determine the fraction of the total radiation from A_1 that is intercepted by A_2.
Solution:
A_1 is a diffuse emitter.

$$E_1 = \pi I \tag{2.14}$$

$$I = E_1/\pi = 9 \times 10^4/\pi = 2.86 \times 10^4 \text{ W/m}^2 sr \tag{2.15}$$

Fig. 2.12 Geometry for
Example 2.1

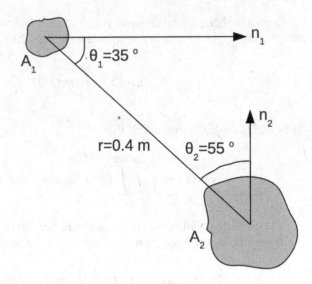

$$I = dQ/(dA_1 . \cos\theta . d\omega) \tag{2.16}$$

$$d\omega = A_2 \cos\theta_2/r^2 = (7 \times 10^{-4} . \cos 55)/0.4^2$$

$$= 2.509 \times 10^{-3} \tag{2.17}$$

$$Q = I.dA_1 . \cos\theta . d\omega \tag{2.18}$$

$$Q = 2.86 \times 10^4 \times 2 \times 10^{-4} \times \cos 35 \times 2.509 \times 10^{-3}$$

$$= 11.756 \times 10^{-3} \text{ W} \tag{2.19}$$

$$Q_{1(\text{total})} = E_1 A_1 = 9 \times 10^4 \times 2 \times 10^{-4} = 18 \text{ W} \tag{2.20}$$

The fraction of the total radiation from A_1 that is intercepted by A_2,

$$Q/Q_1(\text{total}) = 11.756 \times 10^{-3}/18 = 6.53 \times 10^{-4} \tag{2.21}$$

Example 2.2 What is the fraction of the total hemispherical emissive power (E in W/m^2) leaving a diffuse emitter in the direction $20 \leq \theta \leq 50°$ and $10 \leq \phi \leq 70°$.
Solution:

$$E = \int_0^\infty \int_0^{2\pi} \int_0^{\pi/2} I_{\lambda,e} \cos\theta \sin\theta d\theta d\phi d\lambda \tag{2.22}$$

Since $I_{\lambda,e} \neq f(\theta, \phi)$, it can be pulled out of the integral.

$$E = \pi I_b \tag{2.23}$$

$$E_{\text{fraction}} = \int_0^\infty \int_{20}^{50°} \int_{10}^{70°} I_{\lambda,e} \cos\theta \sin\theta\, d\theta\, d\phi\, d\lambda \tag{2.24}$$

$$E_{\text{fraction}} = I_b \times \frac{\pi}{3} \int_{20}^{50°} \cos\theta \sin\theta\, d\theta \tag{2.25}$$

$$E_{\text{fraction}} = I_b \times \frac{\pi}{12} \times 0.9396 \tag{2.26}$$

The fraction of the total radiation is given by

$$\frac{E_{\text{fraction}}}{E_b} = \frac{0.9396\pi I_b}{12\pi I_b} = 0.0783 \tag{2.27}$$

The most important point we have to recognize is that out of 0 to 90° that θ normally varies between, here the θ given is 20–50°, which is 1/3 of the total. And if we consider ϕ, out of 0 to 2π, here it is 10–70°, which is 1/6th of the total. What is the product of these two? It is 1/18 which is 0.0557. But from Eq. (2.27),the solution is not 0.0557 but 0.0783. So there is no shortcut for determining the radiation going out and performing the integration with respect to the angles is inevitable. The problem is simple to work out if the radiator is azimuthally independent. However, zenith angle dependence (i.e. dependence on θ) is not uncommon.

Let us now explore the question "What does the quantity I mean". If the body is diffuse and I is given, one can determine E in the above way or if E is given, one can find out what is the flux coming onto the second surface if we know what is being emitted from the first surface and so on. But suppose we know only the temperature of the first surface, how one can get E or I? This comes from basic radiation laws and the first important concept to know is that radiation also exerts pressure.

2.10 Radiation Pressure

The radiation from the sun exerts a force of nearly 1.2×10^6 kN on the earth. Notwithstanding this, the ability of radiation to exert pressure was unknown till almost the end of the nineteenth century. It was the Italian physicist A. Bartoli who first proposed a thought experiment (whose top view is given in Fig. 2.13) to prove the existence of radiation pressure. The arrangement consists of a cylindrical chamber that is insulated on its periphery so that there is no convection. The cylinder is attached to a black body on the left hand side, which is at temperature T_1 and is attached to another black body on the right hand side which is at temperature T_2 such that $T_1 < T_2$. There are 2 movable frictionless, massless pistons (A and B) contained in this cylindrical enclosure, with a small gap in between, which can be covered or opened by sliding valves. The pistons are perfectly reflecting. Bartoli postulated a thermodynamic cycle with three processes using this.

Initially, the valve which is close to piston 1 is open. The other valve is closed. The region to the left of A, as well as that between A and B, is in contact with a black

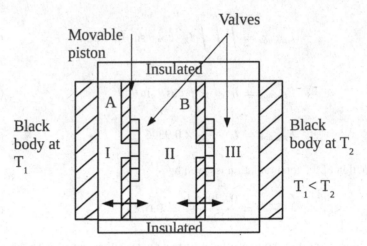

Fig. 2.13 Apparatus for Bartoli's thought experiment

body at temperature T_1. If we give enough time, the cavity consisting of chambers I and II, which is extending up to piston 2, is filled with radiation energy density corresponding to the black body at T_1.

Process 1: The valve near piston A is closed, piston A moves towards piston B till such time that the radiation energy density between the 2 pistons rises to a value corresponding to the black body temperature T_2. This process is akin to compressing a gas. Chamber III is any way in contact with the black body at T_2. Hence, the radiation energy density corresponding to chamber III will be equal to that of the black body at temperature T_2.

Process 2: The piston is opened. The radiation energy densities, on either side of piston 2, are the same as that of the black body at T_2. Both the pistons are moved towards the black body at T_2 such that the energy density is pushed towards the black body at T_2.

Process 3: Bring back the pistons and valves to their initial positions.

In this hypothetical cycle, everything is alright except that the second law of thermodynamics seems to have been violated when heat was transferred from a body at a lower temperature to a body at a higher temperature. The heat could be radiation, conduction or convection and in this case, it happened to be radiation. This is allowed only when extra work is done, and hence we have to find a mechanism or a place where this extra work could have been done. This work could have been done when pushing the pistons towards the right but the pistons are massless. If friction were present, the movement of the pistons would have generated heat which would have increased the temperature. However, the pistons and valves are frictionless. Even so if heat transfer is to take place from a body at a lower temperature T_1 to a body at a higher temperature T_2, some work needs to have been done. Bartoli argued that if everything is correct and the second law of thermodynamics has to be obeyed, this work must have been done against the radiation pressure! Therefore, radiation can be treated like a gas. When we are moving it from T_1 to T_2, we do work and compress it. This is the only way by which we can make the cycle work. Hence, radiation pressure has to exist!

Radiation pressure is not fiction or fantasy and people have measured it using large sails on boats by cutting out the wind component. That said it is quite small and difficult to measure. The existence of radiation pressure opens up new vistas for doing research and exploring "I" because basic thermodynamic laws can now be used to obtain a quantitative handle on radiation. If radiation has pressure and if we also assume that it has internal energy density and if we are able to relate these quantities to I, we can use the TdS (often punned jocularly as "tedious") relations of the classical thermodynamics to relate I and T.

The radiant power passing through dA in Fig 2.11 in the (θ,ϕ) direction in W is

$$dQ(\lambda, \theta, \phi) = I_\lambda(\lambda, \theta, \phi) dA_1 \cos\theta d\omega d\lambda \qquad (2.28)$$

The net momentum flux passing through dA_1 in the (θ, ϕ) direction is tricky to answer. We can relate the momentum to the pressure. But we should be in a position to relate the radiant power to the net momentum flux. How can we do this?

Radiant power / (Area x speed) at which radiant power propagates will have the units $W/(m^2)(m/s) = W\,s/m^3 = Joules/m^3 = Nm/m^3 = N/m^2$ which has the units of pressure.

$$\therefore d\xi(\lambda, \theta, \phi) = \frac{dQ(\lambda, \theta, \phi)}{c.dA} \qquad (2.29)$$

where $d\xi$ has units of N/m^2. Now, the component normal to the radiation will be the component normal to dA.

$$d\xi_n(\lambda, \theta, \phi) = \frac{dQ \cos\theta}{c.dA} \qquad (2.30)$$

Substituting from Eq. 2.27

$$d\xi_n(\lambda, \theta, \phi) = \frac{I_\lambda(\lambda, \theta, \phi)}{c} \cos^2\theta d\omega d\lambda \qquad (2.31)$$

The net change in momentum is equal to the difference between what is going out and what is coming in. Now, if dA is a perfect reflector, whatever is going in must equal whatever is coming out. Therefore, the net change in the flux will be twice the value which is obtained in Eq. (2.31). Now integrating over θ, ϕ and λ, we have the following expression for pressure, P

$$P = \frac{2}{c} \int_0^\infty \int_0^{2\pi} \int_0^{\pi/2} I_{\lambda,e} \cos^2\theta \sin\theta d\theta d\phi d\lambda \qquad (2.32)$$

$$P = \frac{2}{c} I_b 2\pi \int_0^{\pi/2} \cos^2\theta \sin\theta d\theta \qquad (2.33)$$

Between the above two Eqs. (2.32) and (2.33), we have done several things. We have gotten rid of integration with respect to λ by saying that $I_\lambda d\lambda$ can be replaced by I_b. I_b is not a function of (θ, ϕ) because we are assuming everything is a black body so that I_b can be pulled out of the integration. Secondly, we already did the integration with respect to $d\phi$ and have brought the 2π outside the integral. Upon evaluating Eq. (2.33), which turns out to be 1/3, we have

$$P = \frac{4\pi I_b}{3c} \qquad (2.34)$$

This is an important step that thermodynamicists achieved 150 years back when they related the radiation pressure to the intensity.

Next, we define an expression for the **radiation energy density u (J/m^3)**. While radiation energy density has the units J/m^3, radiation flux or power is just W or W/m^2. So, if we have to convert the flux into energy density, the additional variable that enters the problem is time. Therefore, we have to consider a time dt such that in this interval, a length dL is swept by the incident beam and the question we have before us is, if radiation is shining on an object, how much of energy will it accumulate in this dt? dL is the swept length in an interval of time dt. dL can be written as cdt, as shown in Fig. 2.14. The swept volume, dV is given by dL. dA cosθ.

The total amount of energy contained in this volume is

$$dE_\lambda(\lambda, \theta, \phi) = I_{\lambda,e} dA \cos\theta d\omega d\lambda dt \qquad (2.35)$$

Let us now define a du_λ as follows

$$du_\lambda(\lambda, \theta, \phi) = \frac{dE_\lambda(\lambda, \theta, \phi)}{dV} \qquad (2.36)$$

Substituting for dE_λ from Eq. 2.35

$$du_\lambda(\lambda, \theta, \phi) = \frac{I_{\lambda,e}(\lambda, \theta, \phi) dA \cos\theta d\omega d\lambda dt}{c\, dt\, dA\, \cos\theta} \qquad (2.37)$$

Upon integrating from λ = 0 to λ = ∞

$$du(\theta, \phi) = \frac{1}{c} \int_{\lambda=0}^\infty I_{\lambda,e}\, d\omega\, d\lambda \qquad (2.38)$$

$$u = \frac{1}{c} \int I(\theta, \phi) d\omega \qquad (2.39)$$

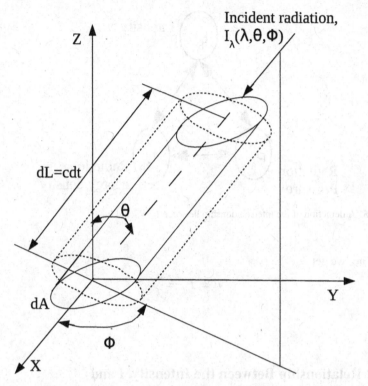

Fig. 2.14 Figure showing a beam of intensity $I_\lambda(\lambda, \theta, \phi)$ incident on the surface element dA

For a black body, $I_b \neq f(\theta, \phi)$ and so $I(\theta, \lambda) = I_b$. The integration is over 4π. Therefore

$$u = u_b = (4\pi I_b)/c \qquad (2.40)$$

The progress we made thus far can be summarized as

(1) an expression for the radiation pressure in terms of the intensity has been derived.
(2) an expression for the radiant energy density in terms of the intensity has been derived.
(3) radiation pressure is basically the radiation intensity divided by speed.

Figure 2.15 gives a depiction of the three central "characters" in the development thus far, namely I_b, P_b and u_b.

From the expressions derived thus far, we have

$$P_b = \frac{4\pi I_b}{3c} \qquad (2.41)$$

$$u_b = \frac{4\pi I_b}{c} \qquad (2.42)$$

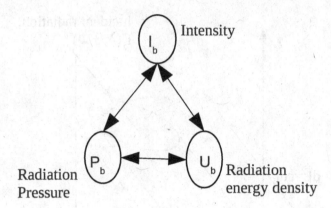

Fig. 2.15 A depiction of the interrelationship between I_b, P_b and u_b

Therefore, we get

$$P = P_b = \frac{u}{3}$$

(2.43)

2.11   Relationship Between the Intensity, I and Temperature, T

When the emissive power of a black body is integrated for all the wavelengths and over the hemisphere, we know that the relationship between E and I is given by

$$E = \pi I_b$$

(2.44)

So getting a relationship between I_b and T is as good as getting a relationship between E and T. This relationship between I_b and T has been independently confirmed by experiments. The question before us is: By thermodynamic arguments, can we get the same relationship?

Consider a gas occupying a volume V, with pressure P, at temperature T.

$$U = uV$$

(2.45)

The total internal energy U is equal to the product of the specific internal energy and the volume. The specific internal energy is defined as energy per unit volume. From the TdS relations in thermodynamics, we have

$$T dS = dU + p dV$$

(2.46)

$$T dS = d(uV) + P dV$$

(2.47)

$$T\,dS = u\,dV + V\,du + P\,dV \tag{2.48}$$

But $P = \dfrac{u}{3}$

$$T\,dS = u\,dV + V\,du + \frac{u}{3}dV \tag{2.49}$$

$$T\,dS = V\,du + \frac{4}{3}u\,dV \tag{2.50}$$

$$dS = \frac{V}{T}du + \frac{4}{3T}u\,dV \tag{2.51}$$

The quantity S, here is entropy, a property and hence it is a point function.
As a consequence, dS is an exact differential.
If

$$Z = f(x, y) \tag{2.52}$$

$$dZ = M\,dx + N\,dy \tag{2.53}$$

For dz to be an exact differential

$$\frac{\partial M}{\partial y} = \frac{\partial N}{\partial y} \tag{2.54}$$

Applying the above to Eq. 2.51 we have

$$\frac{\partial}{\partial V}\left[\frac{V}{T}\frac{\partial u}{\partial T}\right] = \frac{\partial}{\partial T}\frac{4u}{3T} \tag{2.55}$$

$$\frac{1}{T}\frac{du}{dT} = \frac{4}{3T}\frac{du}{dT} - \frac{4u}{3T^2} \tag{2.56}$$

$$\frac{4}{3}\frac{u}{T^2} = \frac{1}{3T}\frac{du}{dT} \tag{2.57}$$

$$\frac{du}{u} = 4\frac{dT}{T} \tag{2.58}$$

Integrating both sides

$$ln\,u = 4ln\,T + ln\,c_1 \tag{2.59}$$

$$u = aT^4 \tag{2.60}$$

Substituting for u from Eq. 2.42

$$\frac{4\pi I_b}{c} = aT^4 \tag{2.61}$$

$$I_b = \frac{ac}{4\pi}T^4 \tag{2.62}$$

For a black body

$$E_b = \pi I_b \tag{2.63}$$

$$\therefore \ E_b = \frac{ac}{4} T^4 \tag{2.64}$$

Equation (2.64) is a very important relationship which was figured out in the last part of the nineteenth century. Stefan and Boltzmann arrived at this independently. In the above equation, **c** is the velocity of light in vacuum $= 2.998 \times 10^8$ m/s. However, the constant a is not known. So, to get the value of a, this expression has to be matched with the values got through experiments. By doing this, it is now known that **ac/4 = 5.67 x 10^{-8} W/m^{2}K^4 $= \sigma$, which is known as Stefan–Boltzmann constant**

$$\therefore \ E_b = \sigma T^4 \tag{2.65}$$

A word about the two scientists. Josef Stefan (1835–1896) and Ludwig Boltzmann (1844–1906) were the two physicists behind this equation. Stefan, an Austrian Professor, was the research supervisor of Boltzmann. It is worth noting that Boltzmann got his PhD at the age of 22 and at the age of 25, he was appointed full professor in mathematical physics at the University of Graz!

Getting back to the distribution, we still do not know what "I" is. But with the help of thermodynamics, without knowing I, we were able to establish that the black body radiation is proportional to the fourth power of temperature. Many questions are still unanswered such as, for a given temperature, how does I_b vary with λ ? Does it hit a peak or are there multiple peaks? For a given wavelength, how does I_b vary with temperature? What happens when λ tends to 0 or when it tends to infinity? The best brains in the last part of the nineteenth century and the early part of the twentieth century worked on this problem and came out with different proposals for $I_b(\lambda)$. These are called **candidate black body distribution functions.**

Intuition suggests that the following constraints have to be satisfied by all possible candidate black body distribution functions $I_b(\lambda)$.

$$I_b(\lambda) \rightarrow 0, \lambda \rightarrow \infty \tag{2.66}$$

$$I_b(\lambda) \rightarrow 0, \lambda \rightarrow 0 \tag{2.67}$$

$$I_b(\lambda) \rightarrow 0, T \rightarrow 0 \tag{2.68}$$

$$I_b(\lambda) \rightarrow \infty, T \rightarrow \infty \tag{2.69}$$

Furthermore, the correct distribution of $I_{b,\lambda}(\lambda, T)$ should also satisfy the following relation

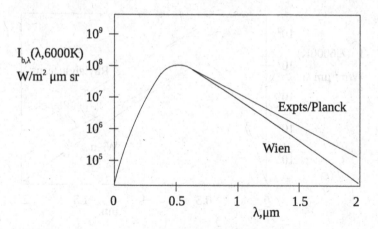

Fig. 2.16 Wien's distribution and comparison with experiments

$$\int_0^\infty \int_0^{2\pi} \int_0^{\pi/2} I_{b,\lambda}(\lambda, \theta, \phi) \ \sin\theta \cos\theta d\theta d\phi d\lambda = \sigma T^4 \tag{2.70}$$

Wien (1864–1928), a German, was the first scientist who proposed a distribution in 1886, as follows

$$I_{b,\lambda}(\lambda, T) = \frac{c_1 \lambda^{-5}}{e^{c_2/\lambda T}} \tag{2.71}$$

Where c_1 and c_2 are the two radiation constants whose values are

$$c_1 = 1.19 \times 10^8 \ W\mu m^4/m^2 \tag{2.72}$$

$$c_2 = 14388 \ \mu mK \tag{2.73}$$

For a black body at 6000 K, while the Wien's distribution holds good for short wavelengths, it deviates from the experimentally obtained distribution at higher values of λ (see Fig. 2.16). Furthermore, when T$\rightarrow \infty$, $e^{\frac{c_2}{\lambda T}} \rightarrow 1$ and so, the curve saturates violating the constraint given by Eq. (2.69). There is a significant departure from the experimental values at higher values of λ. What is special about this temperature ? The temperature under consideration, 6000 K, is important for engineers because it is the temperature of the sun! Therefore, if we take the spectral distribution of $I_{b,\lambda}$ versus λ, corresponding to a black body at 6000 K, the Wien's distribution departs significantly beyond the peak corresponding to the 6000 K. On the left side of the spectrum, though, the distribution does agree with the experiments. It is semi empirical and based on thermodynamics and Wien does not seem to have considered the experimental results available in literature while arriving at the distribution.

The second distribution was proposed by the two scientists Lord Rayleigh (1842–1919) in 1900 and Sir Jeans (1877–1946) in 1905. They both independently worked out yet another incorrect distribution for $I_{b,\lambda}$, as follows

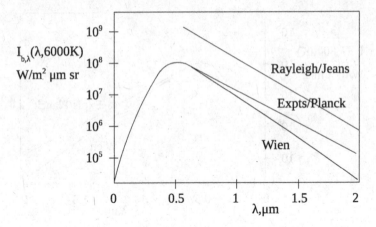

Fig. 2.17 Comparison of candidate distributions with experimental results

$$I_{b,\lambda}(\lambda, T) = \frac{c_1 \lambda^{-5}}{\frac{c_2}{\lambda T}} \qquad (2.74)$$

Let us now see how this differs from the Wien's distribution. The numerator is the same. c_1 and c_2 are the first and second radiation constants, respectively, and have the same value as in the Wien's distribution. The denominator, though, is different. While in the Wien's distribution it is $e^{c_2/\lambda T}$, in the Rayleigh and Jeans distribution, it was $c_2/\lambda T$. The Wien and Rayleigh–Jeans distribution together with that obtained by experiments are shown in Fig. 2.17. At shorter and shorter wavelengths for the distribution proposed by Rayleigh and Jeans, $I_{b,\lambda}(\lambda, T) \to \infty$! But that was not observed by anybody. This means that with very small wavelengths, extremely high $I_{b,\lambda}$ can be produced. This is far from being true. Though the two were very celebrated scientists, their distribution failed miserably at very low wavelengths. So at ultraviolet wavelengths, there is a significant deviation from measured values and this dramatic failure of classical physics in the hands of two greatest physicists was called as the **ultraviolet catastrophe!**

2.12 Planck's Distribution

In 1901 (four years before the publication of the incorrect Rayleigh–Jeans distribution), Max Planck proposed a distribution for $I_{b,\lambda}$ as

$$I_{b,\lambda}(\lambda, T) = \frac{c_1 \lambda^{-5}}{e^{c_2/\lambda T} - 1} \qquad (2.75)$$

where he just added a "-1" in the denominator of the Wien's distribution.

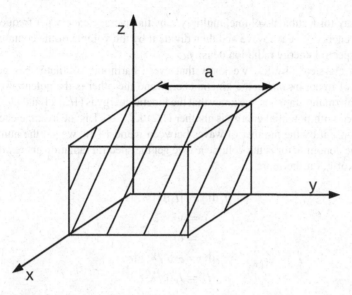

Fig. 2.18 Radiation field enclosed in a cube

His distribution agreed excellently well with the measured $I_{b,\lambda}$ data, at any temperature for all the wavelengths (See Fig. 2.17). Then he started thinking as to why it agreed so perfectly. First, he just did curve fitting. The intriguing part, though, is the fact that Max Planck gave the correct distribution in 1901, while Jeans gave his incorrect distribution four years later (i.e. in 1905). Ironically both papers were published in the same journal.

Planck then wondered about the physics behind the "-1" in the denominator of Eq. 2.75, and figured out that this would not be possible if he adopted the classical physics route. Therefore, Planck figured out that unless he used $E = h\nu$, one will not get the final result. Therefore, he concluded that $E = h\nu$ must be correct. This, in turn, changed the notion that continuous transfer of energy was alone possible and the conclusion was that energy is transferred only in steps of $h\nu$ or in quantum steps of $h\nu$, which in turn, makes it conditional that h have a finite value, which was found to be 6.627×10^{-34} Js now called the **Planck's constant**.

2.13 The Rayleigh–Jeans Distribution

Let us look at the radiation field enclosed in a cubical box of side a and volume a^3 (see Fig. 2.18).

The goal is to determine the radiation energy density within the box, u_ν to further get the relation between u_ν and I_ν. Consider standing waves with a frequency, ν. Between ν and $\nu+d\nu$, if we get the possible number of modes of vibrations or standing waves, say for a cubical volume, multiply it by the average energy per frequency or average energy for each wave and then divide it by the volume of the container, we get the spectral energy radiation density.

From classical physics, we know that every harmonic oscillator has got two degrees of freedom, one is the kinetic energy and the other is the potential energy. From Boltzmann statistics, we know that the kinetic energy is $(1/2)$ kT and the energy associated with potential energy is another $(1/2)$'kT. So kT is the average energy. If we multiply it by the number of waves between v and $v+dv$, we get the numerator and if the denominator is the volume a^3, we get the spectral radiation energy density in terms of v. Furthermore

$$|I_v dv| = |I_\lambda d\lambda| \tag{2.76}$$
$$c = v\lambda \tag{2.77}$$
$$v = c/\lambda \tag{2.78}$$
$$dv = -(c/\lambda^2)d\lambda \tag{2.79}$$
$$I_\lambda = I_v(c/\lambda^2) \tag{2.80}$$

This is the relationship between I_v and I_λ. So, once we have an expression for u_v, we can obtain one for I_v and using Eq. (2.70), in turn, get an expression for $I_{b,\lambda}$. We will now formally derive the Rayleigh–Jeans distribution.

Consider the radiation field enclosed in a cubical box, as shown in Fig. 2.18, whose walls are impermeable. Stationary waves or standing waves are set up inside, whose allowable frequencies have to be determined.

$$u_v = \frac{(\text{No. of standing waves between } (v \text{ and } v + dv) \times (\frac{\text{average energy}}{\text{wave}}))}{a^3} \tag{2.81}$$

$$u_v = \frac{(\text{No. of standing waves between } (v \text{ and } v + dv) \times (kT))}{a^3} \tag{2.82}$$

Mathematical development:

∴ The average energy/wave = kT = potential energy $(1/2 \text{ kT})$ + kinetic energy $(1/2 \text{ kT})$ (where k is the Boltzmann constant given by 1.3806×10^{-23} J/K)

The governing equation for standing waves in the domain is given by the following equation

$$\frac{1}{C^2}\frac{\partial^2\psi}{\partial t^2} = \frac{\partial^2\psi}{\partial x^2} + \frac{\partial^2\psi}{\partial y^2} + \frac{\partial^2\psi}{\partial z^2} \tag{2.83}$$

ψ is the wave function in the Eq. 2.83. For an electromagnetic wave, ψ represents the magnitude of the electric or magnetic field. We are not going to get into the full solution of this equation but have a limited objective of pulling out the number of waves of a prescribed frequency from this. The radiation energy density should be independent of the size of this container. So, it is intuitively apparent that the term a^3

will be present in this expression for the number of waves, so that the denominator term a^3 will see to it that u_ν is independent of a.

Equation (2.83) is a hyperbolic linear partial differential equation, which can be solved using the method of separation of variables. The first step here is to assume a product solution.

$$\psi = T(t)X(x)Y(y)Z(z) \qquad (2.84)$$

We require 2 conditions in time and 6 conditions in space for the cubical container geometry.

$$\psi = 0 \text{ for } x = 0 \text{ and } x = a \qquad (2.85)$$
$$\psi = 0 \text{ for } y = 0 \text{ and } y = a \qquad (2.86)$$
$$\psi = 0 \text{ for } z = 0 \text{ and } z = a \qquad (2.87)$$

The general solution to Eq. (2.83) is given by

$$\psi_n = (A_n \cos \omega_n t + B_n \sin \omega_n t) \sin \frac{n_1 \pi x}{a} \sin \frac{n_2 \pi y}{a} \sin \frac{n_3 \pi z}{a} \qquad (2.88)$$

where n_1, n_2 and n_3 are integers and ω_n is the circular frequency. The circular frequency, ω_n is given by

$$\omega_n^2 = \frac{\pi^2 c^2}{a^2}(n_1^2 + n_2^2 + n_3^2) \qquad (2.89)$$
$$\omega = 2\pi \nu \qquad (2.90)$$
$$\nu_n = \frac{c}{2a}\sqrt{(n_1^2 + n_2^2 + n_3^2)} \qquad (2.91)$$

The challenge is to come up with a number for the discrete frequency modes that are allowed in the frequency interval ν to $\nu+d\nu$. Once we have this, we are kind of done. In order to comprehend its derivation better, let us take recourse to geometry.

Consider a situation shown in Fig. 2.19, where n_1, n_2 and n_3 can take on integer values and all have to be positive. Let us consider a two-dimensional situation, where we have unit squares. We now consider differential area $2\pi r\,dr$ in the first quadrant. What is seen here is a two-dimensional representation using n_1 and n_2 instead of the three-dimensional representation. The hatched area in the figure is given by $2\pi r.dr/4$ (the division by 4 comes in as we are considering only one quadrant). Here $dr = 1$ and the hatched area = 12.6. Now we count the number of lattice points that this hatched area cuts. It is 13. So, if r is sufficiently greater than 1, the area of the quadrant is equal to this number. But, what we want for the spectral density is the number of waves. The number of waves can be related to the area. Here, we have only n_1 and n_2, but in the actual case, we will have n_1, n_2 and n_3 and instead of the quadrant, we will have the first octant of a sphere. So the volume for the octant will be $4\pi r^2 dr/8$ which will be exactly equal to the number of lattice points, which will, in turn, be the number of waves that are allowed between the frequencies ν to $\nu+d\nu$. Let us now try

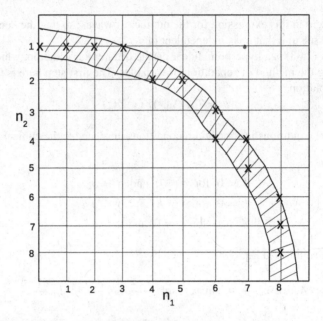

Fig. 2.19 Figure depicting counting in two dimensions

to think of the discrete frequencies in terms of a space lattice, then count the number of lattice points having unit dimensions. For a three-dimensional case, radius r_n is given by

$$r_n = \sqrt{(n_1^2 + n_2^2 + n_3^2)} \tag{2.92}$$

The number of points having a distance between r and r+dr is given by $dN = 4\pi r^2 dr/8$. In a two-dimensional case, the number of points between r and r+dr can be easily counted by not doing a number count but by just taking $2\pi r dr/4$. Therefore, the number of points lying between r and $r + dr$ in a three-dimensional lattice space is given by $4\pi r^2 dr/8$. So the number count was analogous to the elemental area in the case of two dimensions, while it is analogous to the elemental volume in the case of three dimensions. So, we have the following

$$\text{when } r_n = \frac{1}{\sqrt{(n_1^2 + n_2^2 + n_3^2)}} \tag{2.93}$$

$$dN = \frac{4\pi r^2 dr}{8(1)} \tag{2.94}$$

$$\text{and so when, } r_n = \frac{c}{2a}\sqrt{(n_1^2 + n_2^2 + n_3^2)} \tag{2.95}$$

$$dN_\nu = \frac{4\pi \nu^2 d\nu}{8\frac{c}{2a}^3} \tag{2.96}$$

$$dN_\nu = \frac{4\pi a^3}{c^3} \nu^2 d\nu \tag{2.97}$$

The expression for ν_r then becomes (when $d\nu = 1$) the following

$$u_\nu = \frac{\frac{4\pi a^3 \nu^2}{c^3} kT}{a^3} \tag{2.98}$$

$$u_\nu = \frac{4\pi \nu^2}{c^3} kT \tag{2.99}$$

But two possible directions need to be considered. For an electromagnetic wave, two polarizations namely, vertical polarization and horizontal polarization or $E_{parallel}$ and $E_{perpendicular}$ are possible.

$$\therefore u_\nu = (2)\frac{4\pi \nu^2}{c^3} kT x^2 \tag{2.100}$$

$$u_\nu = \frac{8\pi \nu^2}{c^3} kT \tag{2.101}$$

But we know that

$$u_\nu = \frac{4\pi I_{b,\nu}}{c} \tag{2.102}$$

$$I_{b,\nu} = \frac{2\nu^2}{c^2} kT \tag{2.103}$$

But $I_{b,\lambda} = I_{b,\nu} \frac{c}{\nu^2}$ \hfill (2.104)

$$I_{b,\lambda} = \frac{2\nu^2}{c^2} kT \frac{c}{\nu^2} \tag{2.105}$$

$$I_{b,\lambda} = \frac{c}{\lambda^4} kT \tag{2.106}$$

$$2hc^2 = c_1 \tag{2.107}$$

$$\frac{h}{k}c = c_2 \tag{2.108}$$

$$I_{b,\lambda} = \frac{c_1 \lambda^{-5}}{\frac{c_2}{\lambda T}} \tag{2.109}$$

Fig. 2.20 Depiction of an harmonic oscillator, with the help of a spring-mass analogy

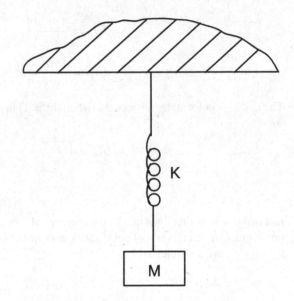

This is the Rayleigh–Jeans distribution. The Rayleigh–Jeans distribution is good for long wavelengths. However, for short wavelengths, it performs very poorly, as already discussed. When $\lambda \rightarrow 0$, $I_{b,\lambda}^{RJ} \rightarrow \infty$, and hence violates Eq. (2.67). This was a dramatic failure of the Maxwell–Boltzmann statistics that was very impressive until then.

2.13.1   Planck's Black Body Distribution Function

The correct distribution was proposed by Planck and for proposing this correct quantum hypothesis, he was awarded the Nobel prize in 1918, at the age of 60. The logic is more or less the same as what Rayleigh–Jean proposed, but there is a modification in terms of the average energy. The number counting is the same as Rayleigh–Jeans. There are three quantities involved, namely, the number of standing waves, the average energy per standing wave and the volume of the container. The volume of the container is the same in both the cases. The number of standing waves is the same. Even so Planck figured out that there is an issue with the value of kT as average energy per standing wave that came from classical physics and kinetic theory of gases. Planck was sure that there was a problem with kT. He used a different approach and used a harmonic oscillator instead to put forth his theory.

    The harmonic oscillator is the equivalent of a spring-mass system as shown in Fig. 2.20, where the stiffness is K and the mass is M. He considered an atomic oscillator as being equivalent to a spring-mass system because here, we can get the natural frequency and other parameters. Two points are to be noted here. (1) The spring-mass oscillator is in thermal equilibrium with its surroundings at a temperature T. In spite of this, it continues to emit radiation, as said by Provost's law and this activity will cease only when the temperature is 0 K (2) The second point is that the oscillator is capable of interacting with electromagnetic radiation.

We will start with classical physics or mechanics and use Boltzmann statistics and determine the total energy, the number of oscillators possible between two energy levels and the total energy of a certain number of oscillators. The total energy divided by the number of oscillators in the energy band will give us the average energy. This average energy should be different from kT. If we happen to get kT again, we will come back to the Rayleigh–Jeans distribution. Planck did some magic there and got values different from kT because of which he got the −1 in the denominator apart from the term got by Wien. The argument goes like this: The proof is correct and finally, one gets an agreement with experiments. If everything is correct, there is a very crucial assumption Planck makes in one step, which must also be correct. Therefore, the theory proposed by Planck is the only one to correctly explain the black body behaviour, as of today. No one else has been able to come with a better explanation that agrees with experimental results.

The total energy of one oscillator with mass M and spring constant K is given by e = KE + PE = $P^2/2M$ + $1/2Kx^2$ where P = instantaneous momentum; x = instantaneous displacement. Any oscillator is characterized by its momentum and displacement.

The number of oscillators having values of (x, P) lying within dx and dP has to come from probability. This comes from the Maxwell–Boltzmann (MB) statistics which actually belong to the pre-quantum era. Planck did not dispute all of what Rayleigh–Jeans said. He used most of the previously available arguments and had problems only with the average energy.

$$dN = NCe^{-e'/kT}dx\,dp \qquad (2.110)$$

The (MB) probability follows an exponential distribution. Equation (2.110) is known as *Arrhenius* type distribution. In (2.110), C is a constant, defined such that $N = \int\int dN$. Somehow, if we know the total number of oscillators at all levels, we can pull out all values except C and get its value.

How will the curves of constant energy appear? This is basically elliptic phase space where the lines are iso energy contours (see Fig. 2.21). There can be several combinations of momentum and displacement that can give the same energy and these when drawn give us elliptical rings, each of a particular energy. We can say one line represents an energy level of e' while the next one is e'+Δe'. Between e' and e'+Δe', if we are able to find out the energy of all the oscillators, we keep it in the numerator. Between e' and e'+Δe', if we find the total number of oscillators and keep it in the denominator of the expression to determine the average energy of the oscillator, then the total energy of all the oscillators between two bands divided by the total number of oscillators between the two bands will give the average energy of any oscillator that lies between the two bands. Once this step is done, we go back to Rayleigh–Jeans and instead of kT, use this value and complete the derivation.

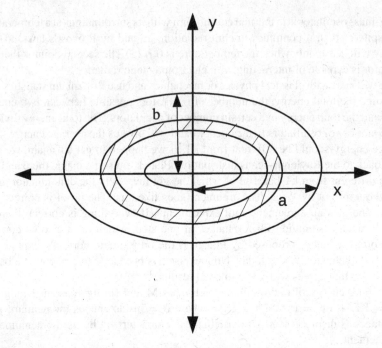

Fig. 2.21 Depiction of the elliptical phase space in the derivation of the Planck's distribution

Let us start by saying that $\Delta e'$ is very small so that $e^{-e'/kT}$ is a constant over $\Delta e'$. The number of oscillators lying in the band e' to $e'+\Delta e'$ is then given by

$$\Delta N = NCe^{-e'/kT} \int\int dxdp \qquad (2.111)$$

What is $\int\int dxdp$? It is the area of the slice of the ellipse indicated by the hatched portion in the Fig. 2.21, denoted by ΔS. The entire area of the ellipse is given by S $= \pi ab$ (a-semi major axis, b-semi minor axis).

In the elliptic phase space, the area of the ellipse is given by $S = \pi\, x_{max}\, P_{max}$.

$$x_{max} = \sqrt{\frac{2e'}{K}} \qquad (2.112)$$

$$P_{max} = \sqrt{2Me'} \qquad (2.113)$$

Now we can get the area S as

$$S = 2\pi e'\sqrt{\frac{M}{K}} \qquad (2.114)$$

The natural frequency of an atomic oscillator with spring constant K and mass M is

$$\nu = \frac{1}{2\pi}\sqrt{\frac{K}{M}} \qquad (2.115)$$

Combining Eqs. 2.114 and 2.115, we have

$$S = \frac{e'}{\nu} \qquad (2.116)$$

$$\Delta S = \frac{\Delta e'}{\nu} \qquad (2.117)$$

Substituting for $\int \int$ dx dp in Eq. 2.111, we have

$$\Delta N = \frac{NCe^{-e'/kT}\Delta e'}{\nu} \qquad (2.118)$$

Recall $\Delta S = \int \int dxdp$.

Now comes the Planck's hypothesis. Consider that the elliptic phase space is divided into increments and into equal bands of area h. Consider a number of such elliptical rings and number them as $n = 0, 1, \ldots$ Planck proposed that the energy of an oscillator located at the inner boundary of a particular ring is e' = $S\nu$ = $nh\nu$, where S is the area between 2 bands that he replaced as nh. If h = 0, we are in trouble as the expression will reduce to kT. So Planck's hypothesis was

$$e' = S\nu = nh\nu \qquad (2.119)$$

There is no rational basis of establishing this by mathematical principles. Also, here, n is a number that can take only integer values.

$$\Delta N = NCe^{(-nh\nu)/kT}\frac{\Delta e'}{\nu} \qquad (2.120)$$

The number of oscillators in a ring "n" is given by

$$N_n = \frac{NC\Delta e'}{\nu}e^{(-nh\nu)/kT} \qquad (2.121)$$

$$N_n = N_0 e^{(-nh\nu)/kT} \qquad (2.122)$$

where

$$N_0 = \frac{NC\Delta e'}{\nu}$$

Now, what is the total energy of all oscillators? The minimum and maximum values of n are 0 and ∞. So we have to add up for all values of n. The total energy of all the oscillators, E, is then given by

$$E = \sum_{n=0}^{\infty} N_n e'_n \tag{2.123}$$

$$E = \sum_{n=0}^{\infty} N_0 e^{-(nh\nu)/kT} .nh\nu \tag{2.124}$$

$$E = N_0 e^{(-nh\nu)/kT} h\nu [1 + 2e^{(-nh\nu)/kT} + 3e^{-(nh\nu)/kT} + \cdots] \tag{2.125}$$

$$\therefore E = \frac{N_0 h\nu e^{-(h\nu)/kT}}{[1 - e^{-(h\nu)/kT}]^2} \tag{2.126}$$

The total number of oscillators N is given by

$$N = \sum_{n=0}^{\infty} N_n \tag{2.127}$$

$$N = N_0 [1 + e^{-(h\nu)/kT} + \cdots] \tag{2.128}$$

$$N = \frac{N_0}{[1 - e^{-h\nu/kT}]} \tag{2.129}$$

We now have the total energy of all the oscillators (2.126) and the total number of oscillators Eq. (2.129). The average energy per oscillator is basically Eq. (2.126) divided by Eq. (2.129), which is

$$W = \frac{N_0 [e^{-h\nu/kT}] h\nu [1 - e^{-h\nu/kT}]}{[1 - e^{-h\nu/kT}]^2 N_0} \tag{2.130}$$

$$W = \frac{h\nu}{(e^{(h\nu/kT)} - 1)} \tag{2.131}$$

Now, Planck wanted to check what will happen when $h\nu/kT$ is very small.

$$\text{When} \qquad \frac{h\nu}{kT} \to 0 \tag{2.132}$$

$$W = \frac{h\nu}{[1 + \frac{h\nu}{kT} - 1]} = kT \tag{2.133}$$

This is what we get in the Rayleigh–Jeans distribution and also Planck was forced to come to the conclusion that h cannot be 0. So h has to be small but finite. So he introduced a fundamental physical constant in nature or in physics which was hitherto unknown and when he matched these results with the experiments, h was found to have a value 6.627×10^{-34} Js. In honour of him, this number is known as **Planck's constant**.

Let us now complete the derivation. The spectral radiational energy density u_ν is given by

$$u_\nu = \frac{4\pi \nu^2}{c^3} \frac{h\nu}{[e^{\frac{h\nu}{kT}} - 1]} \tag{2.134}$$

But $u_\nu = \dfrac{4\pi I_{b,\nu}}{c}$ (2.135)

$$\therefore I_{b,\nu} = \frac{h\nu^3}{c^2[e^{\frac{h\nu}{kT}} - 1]\lambda^2} \tag{2.136}$$

Substituting for ν as $\frac{c}{\lambda}$

$$I_{b,\nu} = \frac{2h(\frac{c}{\lambda})^3}{c^2[e^{\frac{h\nu}{kT}} - 1]\lambda^2} \tag{2.137}$$

$$I_{b,\nu} = \frac{2hc_0^2\lambda^{-5}}{[e^{\frac{hc_0}{\lambda kT}} - 1]} \frac{c_1\lambda^{-5}}{[e^{\frac{c_2}{\lambda T}} - 1]} \tag{2.138}$$

$$c_1 = 2hc_0^2 \tag{2.139}$$

$$c_1 = 2 \times 6.627 \times 10^{-34} \times (2.998 \times 10^8)^2 \tag{2.140}$$

$$c_1 = 1.198 \times 10^{-16} \frac{\text{Jsm}^2}{\text{s}^2} \tag{2.141}$$

$$c_1 = 1.198 \times 10^8 \frac{\text{W}\mu\text{m}^4}{\text{m}^2} \tag{2.142}$$

$$c_2 = \frac{hc_0}{k} = \frac{6.62 \times 10^{-34} \times 2.998 \times 10^8}{1.38 \times 10^{-23}} \tag{2.143}$$

$$c_2 = 14388 \ \mu\text{mK} \tag{2.144}$$

We also know that

$$I_{b,\lambda} = \frac{I_{b,\nu}c}{\nu^2} \tag{2.145}$$

So, finally we have

$$I_{b,\lambda} = \frac{c_1\lambda^{-5}}{[e^{\frac{c_2}{\lambda T}} - 1]} \tag{2.147}$$

c_1 is called the *first radiation constant* and c_2 is called the *second radiation constant*. What is so great about Eq. 2.147? Planck got an expression for $I_{b,\lambda}$, which when plotted against λ for any temperature, gives exactly the same results as what other experimentalists measured for all the wavelengths at all temperatures. Therefore, his result must be correct. If the result is correct, then all the steps he has done to get it must be correct. All, but one of the steps, are exactly the same as what Rayleigh–Jeans have done, that was based on the Maxwell–Boltzmann statistics. There was one crucial departure from the Maxwell–Boltzmann statistics when he proposed that the energy can be divided into bands and $e' = nh\nu$. Therefore, $e' = nh\nu$ must be correct! We get an expression that matches with experimental results only if this assumption is made. **Therefore, the hypothesis that energy transfer must take place only in discrete multiples of $h\nu$ is correct and this was the beginning of quantum mechanics.** He proposed this in 1901, and spent many years after that trying to figure this out. In 1918, he was finally awarded the Nobel prize. Please remember that the integral of $I_{b,\lambda}$ with respect to λ and the solid angle must lead to σT^4.

We will look at Planck's distribution and see if we can extract any further information from that. We have only been seeing the mathematical form. If we plot the distribution versus the wavelength for various temperatures, the Planck's distribution looks like what is shown in Fig. 2.22

$$I_{b,\lambda} = \frac{c_1 \lambda^{-5}}{[e^{\frac{c_2}{\lambda T}} - 1]} \tag{2.148}$$

$$c_1 = 1.198 \times 10^8 \qquad \text{Wmm}^{-4}/\text{m}^2 \tag{2.149}$$

$$c_2 = 1.439 \times 10^4 \qquad \mu\text{mK} \tag{2.150}$$

$$\text{When} \qquad c_2 >> 1 \tag{2.151}$$

$$I_{b,\lambda} = \frac{c_1 \lambda^{-5}}{[e^{\frac{c_2}{\lambda T}}]} = I_{b,\lambda}(\text{Wien}) \tag{2.152}$$

When $c_2 \lambda T >> 1$, the Planck's distribution is approximated to the Wien's distribution, which is valid for very short value of λ or good for short wavelengths. Now, let us see what happens when $\frac{c_2}{\lambda T} \rightarrow 0$

$$I_{b,\lambda} = \frac{c_1 \lambda^{-5}}{[1 + \frac{c_2}{\lambda T} - 1]} = \frac{c_1 \lambda^{-5}}{[\frac{c_2}{\lambda T}]} = I_{b,\lambda}(\text{Rayleigh–Jeans}) \tag{2.153}$$

We see that the Planck's distribution in this case can be approximated to the Rayleigh–Jeans distribution, which is good for long wavelengths, i.e $\frac{c_2}{\lambda T} \rightarrow 0$ $or\, \lambda \rightarrow \infty$. So, the Wien's distribution and the Rayleigh–Jeans distribution are two asymptotes to the Planck's distribution. While the Planck's distribution is valid for all values of the wavelengths, the Wien's distribution and the Rayleigh–Jeans distribution are valid for some portion of the electromagnetic spectrum.

One can now perform a good exercise by taking values of $c_2/\lambda T$ as 10, 100 and 1000 and calculating the value of $I_{b,\lambda}$ using the Planck's distribution, Wien's distribution and the Rayleigh–Jeans distribution and then determining the percentage error. This way we can understand the penalty that we pay for using an approximate expression instead of the correct black body distribution (Planck).

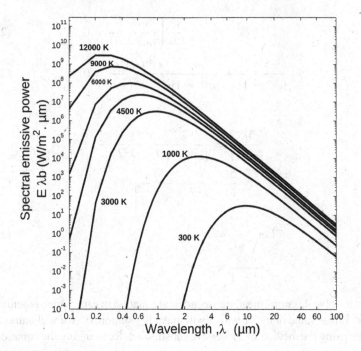

Fig. 2.22 Planck's distribution

2.14 Planck's Distribution—Salient Features

If we look at Fig. 2.22, the following points emerge

1. $I_{b,\lambda}$ is a continuous function of λ. Stated explicitly, for every value of λ, we have a unique value for $I_{b,\lambda}$.
2. For every temperature T, there is a peak.
3. For a given λ, $I_{b,\lambda}$ increases with temperature, as can be seen from the graph. This is intuitively apparent from the second law of thermodynamics, that higher the temperature, higher will be $I_{b,\lambda}$.
4. The peak of the $I_{b,\lambda}$ keeps shifting to the left for increasing temperatures.

How do we get the peak of $I_{b,\lambda}(\lambda, T)$? For this, we do the following

$$\frac{\partial}{\partial \lambda}[I_{b,\lambda}] = 0 \qquad (2.154)$$

We do not want to differentiate $I_{b,\lambda}$ with respect to T because we know that $I_{b,\lambda}$ anyway keeps increasing with T. With λ, though, it increases, reaches a peak and then decreases. Therefore, if $I_{b,\lambda}$ is differentiated with respect to λ and the first derivative equated to 0, it will become stationary and we can get that value of λ that will yield the maximum of $I_{b,\lambda}$.

$$\frac{\partial}{\partial \lambda}\left[\frac{c_1 \lambda^{-5}}{[e^{\frac{c_2}{\lambda T}} - 1]}\right] = 0 \tag{2.155}$$

$$\frac{-5\lambda^{-6}}{[e^{\frac{c_2}{\lambda T}} - 1]} + \frac{-\lambda^{-5}}{[e^{\frac{c_2}{\lambda T}} - 1]^2} \frac{e^{\frac{c_2}{\lambda T}}(-1)c_2}{\lambda^2 T} = 0 \tag{2.156}$$

$$\frac{-5\lambda^{-6}}{[e^{\frac{c_2}{\lambda T}} 1]} = \frac{-\lambda^{-5}}{[e^{\frac{c_2}{\lambda T}} - 1]^2} \frac{e^{\frac{c_2}{\lambda T}}(-1)c_2}{\lambda^2 T}$$

$$\frac{5}{\lambda} = \frac{c_2}{\lambda^2 T} \frac{e^{\frac{c_2}{\lambda T}}}{\left[e^{\frac{c_2}{\lambda T}} - 1\right]} \tag{2.157}$$

$$\text{Let} \quad \frac{c_2}{\lambda T} = x \tag{2.158}$$

$$\frac{xe^x}{e^x - 1} = 5 \tag{2.159}$$

We need to solve this non-linear, transcendental equation to get x which is actually x^* where x^* is that value of $c_2/\lambda T$ which will make $I_{b,\lambda}$ stationary. We will numerically solve this using the method of successive substitution. Rearranging the equation, we get

$$x = \frac{5(e^x - 1)}{e^x} \tag{2.160}$$

$$x_{i+1} = \frac{5(e^{x_i} - 1)}{e^{x_i}} \tag{2.161}$$

We start with some x_i and get x_{i+1}. x_{i+1} is now treated as the new x_i and we keep doing this till the modulus of $(x_{i+1}-x_i)/x_i$ is equal to some acceptably low value. The above is frequently referred to as the convergence criterion or the stopping criterion. Let us start with x = 3 and do the calculations. The results are given in Table 2.1.

One can solve Eq. 2.159 using the Newton–Raphson method too. Here, the advantage is that in the first or second iteration itself, the value of x^* is obtained, as the Newton–Raphson method has quadratic convergence (see Balaji 2019 for a fuller discussion on this method).

Table 2.1 Determination of the root of Eq. 2.161 by successive substitution

Iteration No.	x_i	x_{i+1}	$(x_{i+1}-x_i)^2$
1	3	4.75	3.06
2	4.75	4.95	0.044
3	4.95	4.964	1.96×10^{-4}
4	4.964	4.9654	1×10^{-4}

$$x^* = \frac{c_2}{\lambda T} = 4.965 \quad \mu mK \tag{2.162}$$

$$\lambda_{max} T = \frac{4.1639 \times 10^4}{4.965} \tag{2.163}$$

$$\lambda_{max} T = 2898 \quad \mu mK \tag{2.165}$$

Equation 2.163 is known as the **Wien's displacement law**. We do not require Planck's distribution to get this, by curve fitting too by joining the points of individual maxima of experimentally obtained curves of $I_{b,\lambda}$ Vs T, this can be obtained. Equation 2.165 is a very profound result. Look at the sun's temperature, 6000 K. λ_{max} corresponding to solar radiation is about 0.5 μm. This is so important because 0.5 μm lies in the visible part of the spectrum, which is 0.4–0.7 μm and that is why we have daylight and Earth is so habitable. If the sun's temperature were to be 12000 K, we would require some other source of light throughout the day. The electric bulb, tube light, CFL, LED bulb etc. are all being developed so that the lighting inside mimics the sunlight outside. Our shirt colour and body temperature are at about 37 °C or 300 K. The green colour of the shirt cannot be due to emission as at this temperature, the emission has to be in the infrared part of the spectrum and not in the lease the visible part of the spectrum leave alone the wavelength corresponding to green. From Wien's displacement law, we understand that for all practical purposes, *colour is largely based on reflection rather than emission, unless we encounter high temperatures.*

If we really want to see colours based on emission, we have to take an iron rod and heat it to a high temperature and the colour we see then is because of the emission. So, the key point is that colour can be because of reflection and emission. Suppose we want to selectively absorb radiation, the temperature of the source, is important. If we want to capture the maximum amount of radiation, absorption should be highest in that portion of the spectrum where its maximum lies.

Let us look at something even more interesting. Let us divide the quantity $I_{b,\lambda}$ by the quantity T^5.

$$I_{b,\lambda} = \frac{c_1 \lambda^{-5}}{[e^{\frac{c_2}{\lambda T}} - 1]} \tag{2.166}$$

$$\frac{I_{b,\lambda}}{T^5} = \frac{c_1}{(\lambda T)^5 [e^{\frac{c_2}{\lambda T}} - 1]} = f(\lambda T) \text{ only} \tag{2.167}$$

By innocuously dividing $I_{b,\lambda}$ by a simple quantity such as T^5, we get a very profound result. **The right hand side of (2.167) becomes a function of only λT.** Therefore, we get only one curve as shown in Fig. 2.23, by merging λ and T. The peak corresponds to 2898 μmK. This curve is called the **universal black body distribution function**, whose maximum is the same as that obtained by Wien's displacement law. If we get

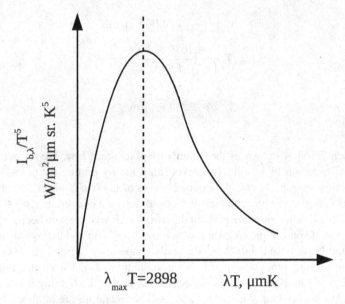

Fig. 2.23 Universal black body distribution function

the area under the curve and treat it to be unity, then between any two wavelengths λ_1 and λ_2, we can find out the fraction that is emitted in a particular band, which is very important.

For example, if we want to design a solar collector, to know the percentage of radiation which is absorbed or emitted in the visible part of the spectrum, the black body radiation function is used and these are called **F function charts.** So, if $I_{b,\lambda}$ is divided by T^5, then the curves get displaced such that only one curve emerges which is the universal black body distribution function. Therefore, some people argue that because some curves get displaced, this should be the Wien's displacement law and not the other one (i.e. Eq. 2.165).

Let us now integrate the Planck's distribution over the hemisphere

$$E_b(\lambda T) = \int I_{b,\lambda} \cos\theta d\omega \tag{2.168}$$

$$E_b(\lambda T) = I_{b,\lambda} \int_0^{2\pi} \int_0^{\pi/2} \cos\theta \sin\theta d\theta d\phi \tag{2.169}$$

$$E_b(\lambda T) = \pi I_{b,\lambda} \tag{2.170}$$

Integrating again from $\lambda = 0$ to ∞

$$E_b(T) = \int_{\lambda=0}^{\infty} \pi I_{b,\lambda} d\lambda \tag{2.171}$$

$$E_b(T) = \pi c_1 \int_{\lambda=0}^{\infty} \frac{\lambda^{-5} d\lambda}{[e^{\frac{c_2}{\lambda T}} - 1]} \tag{2.172}$$

This is a very difficult expression to integrate. We have already introduced the variable $x = c_2/\lambda T$ and using this, we were able to get Wien's displacement law. So, common sense tells us that we can try integrating this too, using the same variable x as before. Please be reminded that this is for a particular temperature.

$$\text{Let } x = \frac{c_2}{\lambda T} \tag{2.173}$$

$$E_b(T) = \pi c_1 \int_0^{\infty} \frac{\lambda^{-5} d\lambda}{[e^x - 1]} \tag{2.174}$$

$$dx = \frac{-c_2}{\lambda^2 T} d\lambda \tag{2.175}$$

$$d\lambda = \frac{-\lambda^2 T dx}{c_2} \tag{2.176}$$

$$E_b(T) = -\pi c_1 \int_{\infty}^{0} \frac{\lambda^{-5} \lambda^2 T}{c_2[e^x - 1]} \tag{2.177}$$

$$E_b(T) = \frac{\pi c_1}{c_2} \int_0^{\infty} \frac{1}{\lambda^3} \frac{T}{(e^x - 1)} dx \tag{2.178}$$

$$E_b(T) = \frac{\pi c_1}{c_2} \int_0^{\infty} \frac{x^3}{\lambda^3} \frac{T^4}{e^x - 1} dx \tag{2.179}$$

$$E_b(T) = \frac{\pi c_1}{c_2^4} T^4 \int_0^{\infty} \frac{x^3 dx}{e^x - 1} \tag{2.180}$$

Now we have to integrate the above expression, which is again not easy. We will use the result that mathematicians have got by integrating such an expression. Please note that T has been taken out of the integral because the integration is with respect to x.

From integral calculus, the following is known

$$\int_0^\infty \frac{x^3 dx}{e^x - 1} = \frac{\pi^4}{15} \tag{2.181}$$

$$\therefore E_b(T) = \frac{\pi c_1}{c_2^4} \frac{\pi^4}{15} T^4 \tag{2.182}$$

$$E_b(T) = 5.67 \times 10^{-8} T^4 \tag{2.183}$$

$$\boxed{E_b(T) = \sigma T^4} \tag{2.185}$$

Planck's law is like the sun of the solar system.

- **For the limit that $c_2/\lambda T$ is very small, we get the Rayleigh–Jeans law**
- **When $c_2/\lambda\, T$ is very large, we get the Wien's law**
- **When we differentiate the Planck's distribution, we get the Wien's displacement law**
- **When we integrate the Planck's distribution, we get the Stefan–Boltzmann law**

As already mentioned, oftentimes we are not interested in finding out the total area under the curve from $\lambda = 0$ to $\lambda = \infty$. Sometimes we may be interested in the amount of energy that is absorbed in the visible part of the spectrum or the infrared part of the spectrum. Or for example, if we have a satellite orbiting the earth in a geostationary orbit at a height of 36000 km and we have an instrument on the satellite that captures the infrared radiation coming from Earth which is altered by the rain, clouds and other particles in the atmosphere (this instrument is a multi-frequency or a multispectral instrument, capable of capturing the radiation at different frequencies). The instrument may have a frequency response in various channels. In each channel, we cannot have a Dirac Delta function. Around a particular $\lambda = 4.5 \ \mu$m, it is not that the radiation captured by it at $\lambda = 4.4 \ \mu$m or $\lambda = 4.6 \ \mu$m is zero. There is no device on earth that can have a frequency response as shown in Fig. 2.24.

Normally, the response of any instrument will be as shown in Fig. 2.25, such that around the λ or ν, there will be a dλ or dν associated. Therefore, there will be a small band of frequencies or wavelengths over which this instrument will respond.

The energy captured will be over the wavelengths λ_1 to λ_2. Hence, the area under the curve between these two will be the energy captured by the instrument. Therefore, now, we are interested in the energy captured in some portions of the spectrum. Therefore, we are also interested in knowing the fraction of the energy which is absorbed, transmitted or reflected between λ_1 and λ_2. Hence, from $E_{b,\lambda}$ versus λ, , if we know λ_1 and λ_2, we are interested in finding out the area of the shaded portion given in Fig. 2.25, for example. So, from 0 to ∞, the total area under the $E_{b,\lambda}$ versus λ curve is σT^4. Out of this, if we know the fraction in λ_1–λ_2, we can multiply this by σT^4 for a body at temperature T and obtain the energy emitted in λ_1–λ_2.

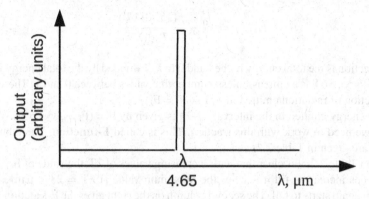

Fig. 2.24 Frequency response curve of a sensor aboard a satellite (ideal)

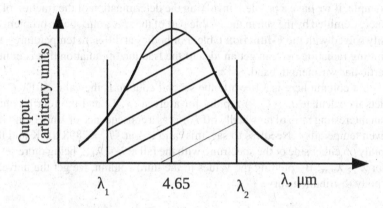

Fig. 2.25 Frequency response curve of a sensor aboard a satellite (actual)

The question before us is what is the fraction of energy emitted by the black body between two limits λ_1 and λ_2 ? Please note that we are not saying T_1 and T_2, because it is for a body at temperature T. The area under the curve between the two wavelengths λ_1 and λ_2 = (area under the curve from $\lambda = 0$ to $\lambda = \lambda_2$)–(area under the curve from $\lambda = 0$ to $\lambda = \lambda_1$)

$$F_{(\lambda_1-\lambda_2)T} \quad or \quad F_{(\lambda_1 T-\lambda_2 T)} \tag{2.186}$$

So, if one can have a look up chart or table that can give us the value of $F_{0-\lambda}$ for any value of λ, then the problem is solved. Suppose we want to find out the area between 0.4 μm and 0.7 μm, then we will take $\lambda_1 = 0.4$ μm and $\lambda_2 = 0.7$μm. If the temperature is known, we will multiply λ and T and first calculate $\lambda_1 T$ and $\lambda_2 T$. Then from the look up chart, using these values of $\lambda_1 T$ and $\lambda_2 T$, we can determine the corresponding fraction and solve the problem easily. Mathematically, this can be written as

$$F_{0-\lambda_1 T} = \frac{\int_0^{\lambda_1 T} f(\lambda T)\mathrm{d}(\lambda T)}{\int_0^{\infty} f(\lambda T)\mathrm{d}(\lambda T)} \tag{2.187}$$

This fraction is the total energy in the band 0 to $\lambda_1 T$ divided by the total energy in the band 0 to ∞. So F is a dimensionless number that varies between 0 and 1. Therefore, the fraction of radiation emitted in $\lambda_1 T - \lambda_2 T = F_{\lambda_1 T - \lambda_2 T}$.

The energy emitted in the interval $\lambda_1 - \lambda_2$ is given by $F = (F_{\lambda_1 T - \lambda_2 T}) \sigma T^4$ W/m^2.

So we need to work with this fraction. This is called **F-function**. The tabulated values are given in Table 2.2.

From Table 2.2, it is clear that with increasing values of λT, the value of $I_{b,\lambda}/\sigma T^5$ also keeps increasing till it reaches the maximum value at $\lambda T = 2898$ μmK, after which it again starts to fall. The second column on the right gives the F-function value or the fraction $F_{0-\lambda T}$. We can use this table intelligently and do a lot of things with it. For example, if we have a problem involving the determination of the fraction of the total energy emitted by the sun in the visible part of the spectrum, such a problem can be easily solved with the **F-function** table. For bodies at different temperatures, that are emitting radiation, we can get an idea of the fraction of radiation that is emitted in a particular wavelength band.

The first column here is λT, while the second column is the value of $F_{0-\lambda}$. The third lets us calculate $I_{b,\lambda}(\lambda, T)$ rapidly for a given (λ, T) and the fourth column gives an interesting ratio of how badly off $I_{b,\lambda}$ is as a consequence of λ not being $\lambda_{\max}$ at a given temperature. Needless to say, this ratio is 1 at $\lambda T = 2898$ μmK and falls off rapidly on either side of the spectrum, with the fall at $\lambda < \lambda_{\max}$ being more severe than for $\lambda > \lambda_{\max}$. By plotting the values in the third column, we get the universal black body distribution curve.

Table 2.2 Blackbody radiation functions

λT (μmK)	$F_{(0 \to \lambda)}$	$I_{\lambda,b}(\lambda, T)/\sigma T^5$ (μm.K.sr)$^{-1}$	$\dfrac{I_{\lambda,b}(\lambda, T)}{I_{\lambda,b}(\lambda_{\max}, T)}$
200	0.000000	3.711772×10^{-28}	0.000000
400	0.000000	4.877254×10^{-14}	0.000000
600	0.000000	1.036654×10^{-9}	0.000014
800	0.000016	9.883195×10^{-8}	0.001370
1000	0.000320	1.182284×10^{-6}	0.016385
1200	0.002130	5.228789×10^{-6}	0.072464
1400	0.007778	1.341736×10^{-5}	0.185946
1600	0.019691	2.487352×10^{-5}	0.344712
1800	0.039292	3.750250×10^{-5}	0.519732
2000	0.066653	4.927725×10^{-5}	0.682914
2200	0.100782	5.889147×10^{-5}	0.816153
2400	0.140119	6.580943×10^{-5}	0.912027

(continued)

Table 2.2 (continued)

λT (μmK)	$F_{(0 \to \lambda)}$	$I_{\lambda,b}(\lambda, T)/\sigma T^5$ (μm.K.sr)$^{-1}$	$\dfrac{I_{\lambda,b}(\lambda, T)}{I_{\lambda,b}(\lambda_{max}, T)}$
2600	0.182951	7.005151×10^{-5}	0.970816
2800	0.227691	7.194804×10^{-5}	0.997099
2898	0.249913	7.215735×10^{-5}	1.000000
3000	0.273004	7.195287×10^{-5}	0.997166
3200	0.317847	7.052908×10^{-5}	0.977434
3400	0.361457	6.809082×10^{-5}	0.943644
3600	0.403307	6.498093×10^{-5}	0.900545
3800	0.443063	6.146872×10^{-5}	0.851870
4000	0.480541	5.775734×10^{-5}	0.800436
4200	0.515662	5.399468×10^{-5}	0.748291
4400	0.548431	5.028467×10^{-5}	0.696875
4600	0.578903	4.669735×10^{-5}	0.647160
4800	0.607171	4.327738×10^{-5}	0.599764
5000	0.633350	4.005083×10^{-5}	0.555048
5200	0.657564	3.703038×10^{-5}	0.513189
5400	0.679946	3.421938×10^{-5}	0.474233
5600	0.700626	3.161478×10^{-5}	0.438137
5800	0.719732	2.920932×10^{-5}	0.404800
6000	0.737386	2.699316×10^{-5}	0.374087
6200	0.753704	2.495496×10^{-5}	0.345841
6400	0.768793	2.308268×10^{-5}	0.319894
6600	0.782754	2.136417×10^{-5}	0.296077
6800	0.795680	1.978750×10^{-5}	0.274227
7000	0.807657	1.834121×10^{-5}	0.254183
7200	0.818763	1.701444×10^{-5}	0.235796
7400	0.829070	1.579705×10^{-5}	0.218925
7600	0.838643	1.467960×10^{-5}	0.203439
7800	0.847543	1.365338×10^{-5}	0.189217
8000	0.855825	1.271039×10^{-5}	0.176148
8200	0.863538	1.184331×10^{-5}	0.164132
8400	0.870728	1.104546×10^{-5}	0.153075
8600	0.877437	1.031075×10^{-5}	0.142893
8800	0.883702	9.633637×10^{-6}	0.133509
9000	0.889559	9.009093×10^{-6}	0.124853
9200	0.895038	8.432545×10^{-6}	0.116863
9400	0.900169	7.899844×10^{-6}	0.109481
9600	0.904977	7.407225×10^{-6}	0.102654

(continued)

Table 2.2 (continued)

λT (μmK)	$F_{(0\to\lambda)}$	$I_{\lambda,b}(\lambda, T)/\sigma T^5$ (μm.K.sr)$^{-1}$	$\dfrac{I_{\lambda,b}(\lambda, T)}{I_{\lambda,b}(\lambda_{max}, T)}$
9800	0.909488	6.951272×10^{-6}	0.096335
10000	0.913723	6.528882×10^{-6}	0.090481
10500	0.923232	5.601903×10^{-6}	0.077635
11000	0.931410	4.830388×10^{-6}	0.066942
11500	0.938479	4.184824×10^{-6}	0.057996
12000	0.944616	3.641843×10^{-6}	0.050471
12500	0.949969	3.182853×10^{-6}	0.044110
13000	0.954656	2.792992×10^{-6}	0.038707
13500	0.958777	2.460320×10^{-6}	0.034097
14000	0.962413	2.175193×10^{-6}	0.030145
14500	0.965634	1.929783×10^{-6}	0.026744
15000	0.968496	1.717707×10^{-6}	0.023805
15500	0.971047	1.533730×10^{-6}	0.021255
16000	0.973328	1.373542×10^{-6}	0.019035
16500	0.975374	1.233578×10^{-6}	0.017096
17000	0.977214	1.110874×10^{-6}	0.015395
17500	0.978873	1.002956×10^{-6}	0.013900
18000	0.980373	9.077520×10^{-7}	0.012580
18500	0.981732	8.235169×10^{-7}	0.011413
19000	0.982966	7.487779×10^{-7}	0.010377
19500	0.984090	6.822862×10^{-7}	0.009456
20000	0.985114	6.229790×10^{-7}	0.008634
25000	0.991726	2.763310×10^{-7}	0.003830
30000	0.994851	1.403976×10^{-7}	0.001946
35000	0.996514	7.862366×10^{-8}	0.001090
40000	0.997478	4.736531×10^{-8}	0.000656
45000	0.998075	3.020096×10^{-8}	0.000419
50000	0.998464	2.015049×10^{-8}	0.000279
55000	0.998728	1.395263×10^{-8}	0.000193
60000	0.998914	9.964006×10^{-9}	0.000138
65000	0.999048	7.303743×10^{-9}	0.000101
70000	0.999148	5.474729×10^{-9}	0.000076
75000	0.999223	4.183928×10^{-9}	0.000058
80000	0.999281	3.252025×10^{-9}	0.000045
85000	0.999327	2.565680×10^{-9}	0.000036
90000	0.999363	2.051193×10^{-9}	0.000028
95000	0.999392	1.659422×10^{-9}	0.000023
100000	0.999415	1.356864×10^{-9}	0.000019

Example 2.3 Consider a black body at a temperature of 6000 K. Determine the following. (a) $I_{b,\lambda}$ at 0.4 μm (b) $I_{b,\lambda}$ at 0.01 μm (c) $I_{b,\lambda}$ at 10 μm (d) total hemispherical emissive power $E_b(T)$ (e) $I_{b,\lambda}$ corresponding to λ_{max} (f) ratio of $I_{b,\lambda}$ at λ_{max} to $I_{b,\lambda}$ at 10 μm (g) fraction of radiation in the visible part of the spectrum.
Solution:

Temperature of the black body T = 6000 K

(a) $I_{b,\lambda}$ at 0.4 μm

$$I_{b,\lambda} = \frac{c_1 \lambda^{-5}}{[e^{\frac{c_2}{\lambda T}} - 1]} \tag{2.188}$$

$$I_{b,\lambda} = \frac{1.198 \times 10^8 (0.4)^{-5}}{[e^{\frac{1.439 \times 10^4}{0.4 \times 6000}} - 1]} \tag{2.189}$$

$$I_{b,\lambda} = 2.91 \times 10^7 \ \text{W/m}^2\text{m m sr} \tag{2.190}$$

(b) $I_{b,\lambda}$ at 0.01 μm = 0 (hardly any radiation in this ultraviolet part of the spectrum from the sun)
(c) $I_{b,\lambda}$ at 10 μm = 4420 W/m²μm sr
(d) $E_b(T) = \sigma T^4 = 7.35 \times 10^7$ W/m²
(e) $\lambda_{max}T = 2898$ μmK; $\lambda_{max} = 0.483$ μm (this corresponds to blue light as the wavelength of blue radiation is between 0.45 μm and 0.49 μm! Of course, if we want to do it for solar radiation, instead of 6000K, we will use 5800 K.) $I_{b,\lambda}$ corresponding to $\lambda_{max} = 3.181 \times 10^7$ W/m²μmsr
(f) $I_{b,\lambda}$ at λ_{max} / $I_{b,\lambda}$ at 10 μm = 7196;
This is a very important result. The $I_{b,\lambda}$ corresponding to visible blue light divided by the $I_{b,\lambda}$ at infrared of 10 μm for the temperature of 6000 K, which corresponds to the outer temperature of the sun, we can see that the ratio of intensity of visible radiation to infrared radiation is about 7196.
(g) Fraction in the visible part of the spectrum: $\lambda_1 = 0.4$ μm, $\lambda_2 = 0.7$μm, T = 6000 K, $\lambda_1 T = 2400$ μmK, $\lambda_2 T = 4200$ μmK

The fraction corresponding to $\lambda_1 T = 2400$ μmK, $F_{0-\lambda_1 T} = 0.14$.
The fraction corresponding to $\lambda_2 T = 4200$ μmK, $F_{0-\lambda_2 T} = 0.516$.
$F_{\lambda_1 - \lambda_2} = 0.376$ or 37.6%.
Please do not underestimate the importance of this result because 0 to ∞ is so huge and the wavelength band of 0.4–0.7 μm is so small, yet the fraction of radiation in this band is almost 40% of the total radiation in the 0–∞ band. So, nearly 40% of the solar radiation is concentrated in a very small portion of the spectrum, namely the visible range. If the temperature of the sun were not 6000 K, the peak would be in some other part of the spectrum and we would not be getting enough "visible" radiation.

So, now the question arises: if we have the temperature of the sun to be 6000 K, which makes the earth so habitable, when all of it started with the big bang theory (assuming this to be the correct theory to explain the origins of the universe), it was basically a cooling problem. The cooling started with some initial conditions. If the initial conditions were to be different, then the temperature of the photosphere of the sun would have been different, which means that the fraction of the radiation falling would have been different and the earth's temperature may have been different because of which we all may not have been here today! Why were the initial conditions chosen that way, or who chose them to be so?

Problems

2.1 (a) Verify that the spectral intensity distribution given by Planck ($I_{b,\lambda}$) when divided by σT^5 becomes a function of λT alone.
  (b) Obtain the value of λT at which the quantity given by (a) becomes the maximum.
  (c) Hence, obtain the maximum value of the quantity $I_{(b,\lambda)}/\sigma T^5$ and verify it with the value given in the F Tables.

2.2 Compute the fraction of total, hemispherical emissive power leaving a diffuse surface in the direction $\frac{\pi}{6} \le \theta \le \frac{\pi}{3}$ and $0 \le \phi \le \frac{5\pi}{6}$.

2.3 The directional, total intensity of solar radiation incident on the surface of the earth on a bright sunny day at a particular location in the tropics is given by $I_\theta = I_n \cos\theta$, where $I_n = 500 \, \text{W/m}^2$ sr is the total intensity of radiation corresponding to $\theta = 0$ (θ is the zenith angle). Determine the solar irradiation at the earth's surface.

2.4 Show that the Planck's distribution reduces to $E_{b,\lambda} = c_1 T/c_2\lambda^4$ when $c_2/\lambda T \ll 1$. Compute the error with respect to the Planck's distribution when $\lambda T = 1.8 \times 10^5 \, \mu\text{mK}$ and comment on your result.

2.5 Show that the Planck's distribution reduces to $E_{b,\lambda} = c_1/\lambda^5 e^{-c_2/\lambda T}$ when $c_2/\lambda T \gg 1$. Compute the error with respect to the Planck's distribution at $\lambda T = 2898 \, \mu\text{mK}$ and comment on your result.

2.6 Determine the sun's radiation intensity at the middle of the visible spectrum assuming that the outer surface of the sun is a black body at 5800 K.

2.7 Estimate the temperature of the earth's surface (assuming it to be black) given that the outer surface of the sun has an equivalent temperature of 5800 K. The diameters of the sun and the earth may be taken to be 1.39×10^9 m and 1.29×10^7 m, respectively, and the distance between the sun and the earth is 1.5×10^{11} m.

2.8 Determine the wavelength corresponding to the maximum emission from each of the following surfaces: the sun, a tungsten filament at 2900 K, a heated metal at 1400 K, earth's surface at 300 K, and a metal surface in outer space at 70 K.

2.9 Using the data given in Problem 2.8, estimate the fraction of the solar emission that is in the following spectral regions: the ultraviolet, the visible and the infrared. Compute these values for the tungsten filament. Compare the fraction of emission of the two sources in the visible part of the spectrum and comment on your findings.

Chapter 3
Radiative Properties of Non-black Surfaces

Thus far, we have considered only a black body. In reality, though, it is almost impossible to encounter a black body. Real bodies are neither perfect absorbers nor perfect emitters. So there is a deviation or departure from black body behaviour. As engineers, we have to live with real surfaces. Once, we know upfront that real surfaces are not black bodies, we need to characterize their behaviour in so far as radiation is concerned. Therefore, we have to introduce the concept of **radiation surface properties**.

The goal of this characterization is to be able to quantify the departure of real bodies from black body behaviour. This departure manifests itself as incomplete absorption and imperfect emission. Consider Fig. 3.1, that presents the typical variation of spectral intensity of emission with wavelength for a black body at 1073 K (about 800 °C), a temperature normally encountered in engineering. Using the Planck's distribution, we can plot the curve of $I_{b,\lambda}$ versus λ and the peak of the distribution is around 2.8 μm, which is consistent with the Wien's displacement law. We can have a body corresponding to curve "a" which is called a **gray body** at 1073 K, whose ratio of emission at a particular wavelength to that by a blackbody is fixed.

Mathematically, for a gray surface,

$$\frac{I_\lambda(\lambda, T)}{I_{b,\lambda}(\lambda, T)} \neq f(\lambda) \tag{3.1}$$

Any body/surface satisfying Eq. (3.1) is known as a gray body/surface.

Why do we need a gray body model?

The gray body is an idealization, which we use because it helps us to simplify calculations in radiative heat transfer. Otherwise, if we want to consider this ratio as a function of λ, the analysis becomes more tedious. Curve "b" (of Fig. 3.1a) is actually more representative of the behaviour of most real surfaces. So we can guess that radiative analysis of a surface which follows "b" is a lot more difficult than the

© The Author(s), under exclusive license to Springer Nature Switzerland AG 2021
C. Balaji, *Essentials of Radiation Heat Transfer*,
https://doi.org/10.1007/978-3-030-62617-4_3

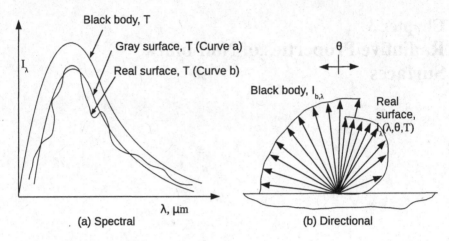

Fig. 3.1 Typical distribution of I_λ for real surfaces

behaviour which follows "a". But still, the area under the curves "a" and "b" may be more or less the same. Even so, if we use the gray body assumption and use the smooth curve, it may lead to some local errors. For example, when we try to define the value of I_λ for a particular value of λ, there may be a noticeable difference between the I_λ for a gray body and that for the real surface. However, when we average out and integrate from 0 to ∞, the error may not be significant.

But why would we want to use this? Because, for most surfaces, I_λ versus λ is not known. The ratio given above is called **emissivity** and when it is a function of wavelength, it is called **spectral emissivity** and this is unknown for most surfaces, and therefore, we go in for gray body behaviour. The other reason is that even if the spectral emissivity is known, often we do not know how to incorporate this information into our analysis, which is generally the case. So, because of these two reasons, namely, either the spectral emissivity is not known or if known, we do not have the competence or wherewithal to use this information, people go in for a simple assumption where $I_\lambda / I_{b,\lambda}$ is not a function of λ. This is called the **gray body assumption**.

Now if we look at Fig. 3.1b, it gives the directional spectral intensity I_λ, at a given wavelength, temperature and azimuthal angle for a black surface, a diffuse surface and a hypothetical real surface. The zenith angle, θ is with respect to the vertical. For purposes of drawing this graph, $T = T_1$, which is fixed. The wavelength $\lambda = \lambda_1$, is again fixed and so is ϕ which takes on the value ϕ_1. There are basically 4 parameters here, namely the wavelength, temperature, zenith angle and the azimuthal angle, where the wavelength, temperature and the azimuthal angle are fixed. We are studying the variation of I_λ with respect to θ alone. We already said that the black body emission is independent of all angles, including the zenith angle. Hence, we get a semicircular shape for I_λ as it varies with θ, as the magnitude remains the same.

Curve "a" corresponds to a surface whose I_λ is not a function of θ. At any θ, it will have a value or magnitude smaller than that of the black body as the black body emission is the maximum possible. Hence, we get a curve that is concentric with the curve obtained for the black body, but which has a smaller magnitude. This leads us to the concept of a diffuse surface.

$$\frac{I_{\lambda,e}(\lambda, T, \theta, \phi)}{I_{b,\lambda}(\lambda, T)} \neq f(\theta) \tag{3.2}$$

Now we can draw one more curve, where, instead of keeping ϕ fixed, we keep θ fixed and say that the azimuthal angle is a variable. Therefore, we should have a general case where, for a diffuse surface,

$$\frac{I_{\lambda,e}(\lambda, T, \theta, \phi)}{I_{b,\lambda}(\lambda, T)} \neq f(\theta, \phi) \tag{3.3}$$

Now if we say that a body is simultaneously gray and diffuse, then for this dimensionless ratio, its functional dependence on λ, θ and ϕ is knocked out. This dimensionless ratio, which is emissivity, becomes a function of only the temperature. Then if we do some engineering analysis, and we are also working in a very narrow temperature range, we can say that the emissivity is not a function of temperature in this case, and thus say that it does not depend on any variable. However, it is instructive to mention here that many approximations are involved in reaching that step and we must be aware of the assumptions we are making.

Why do we use this gray diffuse approximation? The answer is: Many engineering materials conform to this behaviour and it helps us do the radiative transfer calculations very fast. It also helps us combine radiation with convection and conduction easily in multimode problems of engineering interest. Therefore, the gray diffuse approximation is very useful, potent and is frequently used in engineering practice.

3.1 Spectral Directional Emissivity, $\epsilon'_\lambda (\lambda, T, \theta, \phi)$

The spectral directional emissivity given by ϵ'_λ is the ratio of the spectral directional intensity of emission of a real surface to the spectral radiation intensity of a black body at that wavelength, at that temperature in the same direction. Mathematically $\epsilon'_\lambda(\lambda, T, \theta, \phi)$ is given by

$$\epsilon'_\lambda(\lambda, T, \theta, \phi) = \frac{I_{\lambda,e}(\lambda, T, \theta, \phi)}{I_{b,\lambda}(\lambda, T)} \tag{3.4}$$

So, as expected, ϵ'_λ is a dimensionless ratio, which varies between 0 and 1. It is a non-dimensional way of declaring the efficiency of emission of a surface. To say how efficient a surface is, we need a benchmark or a standard, which is the black body here. Corresponding to a black body, how efficiently the surface is emitting is what this number conveys. For a given temperature, wavelength and direction, if ϵ'_λ

is given, either from theory or from experiments, we can use the Planck's distribution and get $I_{b,\lambda}$, multiply these two and get the value of $I_{\lambda,e}$ using the equation given below.

$$I_{\lambda,e}(\lambda, T, \theta, \phi) = \epsilon'_\lambda(\lambda, T, \theta, \phi).I_{b,\lambda}(\lambda, T) \tag{3.5}$$

For a gray body or a gray surface

$$\epsilon'_\lambda(\lambda, T, \theta, \phi) \neq f(\lambda) \tag{3.6}$$

$$\epsilon'_\lambda(\lambda, T, \theta, \phi) = \epsilon'_\lambda(T, \theta, \phi) \neq f(\lambda) \tag{3.7}$$

So for a gray body, ϵ'_λ is independent of λ. Though ϵ'_λ is not a function of λ, it is still a spectral quantity. Unfortunately or fortunately, that value is the same for all the wavelengths. What do we mean by saying that it is a spectral quantity? The integration with respect to λ is not done, so it is spectral. If integration with respect to angle is not done, it is called directional. Even though we removed the functional dependence on λ, we still call it ϵ'_λ. There could be surfaces that need not exhibit gray body behaviour for all angles. For particular zenith and azimuthal angles, they can exhibit gray body behaviour; in other angles, they may not exhibit gray body behaviour. That depends on their nature. If in the angles of our interest, a surface exhibits gray body behaviour, the analysis becomes easier.

For a diffuse surface, the following relations hold:

$$\epsilon'_\lambda(\lambda, T, \theta, \phi) \neq f(\theta, \phi) \tag{3.8}$$

$$\epsilon'_\lambda(\lambda, T, \theta, \phi) = \epsilon'_\lambda(\lambda, T) \tag{3.9}$$

The term "diffuse" is with respect to angle and "gray" is with respect to wavelength. For a gray, diffuse surface, then, we have

$$\epsilon'_\lambda(\lambda, T, \theta, \phi) = f(T) \quad \text{only} \tag{3.10}$$

Equation (3.10) is a very powerful approximation and many surfaces exhibit this behaviour.

There are surfaces where this gray diffuse approximation is not valid. So for such surfaces from ϵ'_λ, we have to get to ϵ. For this, we need to integrate ϵ'_λ, with respect to λ, θ and ϕ. This will be the basic parameter ϵ, which we would have used in a heat transfer course. ϵ is a function of λ for many surfaces.

Consider a surface for which α is the solar absorptivity. If we are interested in a solar collector, the α should be very high. But then α means the absorptivity corresponding to the sun's temperature, 5800 K. The body will get heated but its temperature will be about 70 or 80 °C. However, when it emits, according to Wien's displacement law, I_λ maximum will be around 10 μm. So, for the design of a solar collector, we would want to look at a surface where the emissivity corresponding to infrared is low and the absorptivity corresponding to incoming solar radiation is high.

So if α is high and ϵ is low, it will start collecting heat and it will be like a green house. We would want the exactly opposite behaviour when we want to design a system that cuts out the radiation and keeps the interiors cool. This is what is achieved by sun control films and double paned glasses. So, depending on the application, we can play with or engineer α and ϵ, which are functions of λ. These are called "selective emitters" and "selective absorbers".

3.2 Hemispherical Spectral Emissivity, $\epsilon_\lambda(\lambda, T)$

Mathematically hemispherical spectral emissivity, $\epsilon_\lambda(\lambda, T)$ is defined as

$$\epsilon_\lambda(\lambda, T) = \frac{E_\lambda(\lambda, T)}{E_b(\lambda, T)} \tag{3.11}$$

Let us consider a surface which is at a particular temperature. We are trying to find out the radiation emission from this surface over a hemispherical basket, at each and every wavelength, for which we have to integrate with respect to θ and ϕ. That is why we knocked off the θ and ϕ in the numerator of the expression given above. The numerator of Eq. (3.11) can be obtained as follows:

$$E_\lambda(\lambda, T) = \int_{\phi=0}^{2\pi} \int_{\theta=0}^{\frac{\pi}{2}} I_{\lambda,e}(\lambda, T, \theta, \phi) \cos\theta \sin\theta d\theta d\phi \tag{3.12}$$

However, from the definition of spectral directional emissivity, we have

$$I_{\lambda,e}(\lambda, T, \theta, \phi) = \epsilon_\lambda'(\lambda, T, \theta, \phi) I_{b,\lambda}(\lambda, T) \tag{3.13}$$

$$\therefore E_\lambda(\lambda, T) = \int_{\phi=0}^{2\pi} \int_{\theta=0}^{\frac{\pi}{2}} \epsilon_\lambda'(\lambda, T, \theta, \phi) I_{b,\lambda}(\lambda, T) \cos\theta \sin\theta d\theta d\phi \tag{3.14}$$

We are already able to see some silver lining, as we know that $I_{b,\lambda}$ is not a function of ϕ and θ. Before doing this, we can substitute this expression (Eq. 3.14) in place of E_λ in our original definition of ϵ_λ or Eq. (3.11).

$$\epsilon_\lambda(\lambda, T) = \frac{I_{b,\lambda}(\lambda, T) \int_{\phi=0}^{2\pi} \int_{\theta=0}^{\frac{\pi}{2}} \epsilon_\lambda'(\lambda, T, \theta, \phi) \cos\theta \sin\theta d\theta d\phi}{\pi I_{b,\lambda}(\lambda, T)} \tag{3.15}$$

$$\therefore \epsilon_\lambda(\lambda, T) = \frac{1}{\pi} \int_{\phi=0}^{2\pi} \int_{\theta=0}^{\frac{\pi}{2}} \epsilon_\lambda'(\lambda, T, \theta, \phi) \cos\theta \sin\theta d\theta d\phi \tag{3.16}$$

From this equation, it is clear that if we know ϵ'_λ we can determine ϵ_λ. So, Eq. (3.16) is a powerful expression which relates a spectral, directional quantity to a spectral, hemispherical quantity. The quantity in Eq. (3.16) is hemispherical because we used $\theta = 0$ to $\pi/2$ in the integration and not $-\pi/2$ to $+\pi/2$. We are only looking at radiation from a hemisphere. Equation (3.16) is generic enough that it can be applied to reflectivity, absorptivity and so on. So if we give ϵ'_λ in the form of tables or data sheets, we can integrate and calculate ϵ_λ. From ϵ_λ, if we do one more integration, we will get ϵ. Then we can use the Stefan Boltzmann's law to get, $E = \epsilon\sigma T^4$.

In the above expression $\cos\theta\sin\theta d\theta d\phi$ is basically $d\omega$ together with the idea of projected area and the integral in the numerator turns out to π, should we have a diffuse surface. Hence, for a diffuse surface, the following additional relations hold:

$$\epsilon'_\lambda \neq f(\theta, \phi) \tag{3.17}$$

$$\epsilon_\lambda(\lambda, T) = \frac{\epsilon'_\lambda}{\pi} \int_{\phi=0}^{2\pi} \int_{\theta=0}^{\frac{\pi}{2}} \cos\theta \sin\theta d\theta d\phi = \epsilon'_\lambda(\lambda, T, \theta, \phi) \tag{3.18}$$

3.3 Directional Total Emissivity, $\epsilon'(T, \theta, \phi)$

The directional total emissivity is represented by $\epsilon'(T, \theta, \phi)$. The prime in the symbol denotes that it is still a directional quantity, but we got rid of the λ and hence it is a total emissivity. Mathematically, $\epsilon'(T, \theta, \phi)$ is given by

$$\epsilon'(T, \theta, \phi) = \frac{E'(T, \theta, \phi)}{E'_b(T, \theta, \phi)} \tag{3.19}$$

$$\epsilon'(T, \theta, \phi) = \frac{E'(T, \theta, \phi)}{I'_b(T, \theta, \phi)\cos\theta} \tag{3.20}$$

The numerator in Eq. (3.19) can be written as

$$E'(T, \theta, \phi) = \int_{\lambda=0}^{\infty} E'_\lambda(\lambda, T, \theta, \phi)d\lambda \tag{3.21}$$

The procedure we are following to derive this is to introduce the definition of emissivity formally. The denominator is that of a black body. We only manipulate the numerator. Equation (3.20) is a slightly changed version of Eq. (3.19), where the E_b is changed to I_b. Now, in order to get ϵ', we have to somehow link it to ϵ'_λ, because we know ϵ'_λ. So on the right side, we have to introduce ϵ'_λ. In the numerator, we have replaced the E' by the integral expression which has E'_λ within the integral sign. We know that E'_λ can be written in terms of ϵ'_λ. Replacing E'_λ as $I'_\lambda \cos\theta$ in Eq. (3.21)

$$E'(T, \theta, \phi) = \int_{\lambda=0}^{\infty} I_\lambda'(\lambda, T, \theta, \phi) \cos\theta \, d\lambda \qquad (3.22)$$

Furthermore,

$$I_\lambda'(\lambda, T, \theta, \phi) = \epsilon_\lambda'(\lambda, T, \theta, \phi) I_{b,\lambda}(\lambda, T) \qquad (3.23)$$

$$\therefore E'(T, \theta, \phi) = \int_{\lambda=0}^{\infty} \epsilon_\lambda'(\lambda, T, \theta, \phi) I_{b,\lambda}(\lambda, T) \cos\theta \, d\lambda \qquad (3.24)$$

Now we can substitute for E'(T,θ, ϕ) in the original definition for $\epsilon'(T, \phi, \theta)$, in Eq. (3.19). Cancelling out the $\cos\theta$ in the numerator and the denominator, we get

$$\epsilon'(T, \theta, \phi) = \frac{\int_0^{\infty} \epsilon_\lambda'(\lambda, T, \theta, \phi) I_{b,\lambda}(\lambda, T) d\lambda}{\frac{\sigma T^4}{\pi}} \qquad (3.25)$$

Now, the acid test is, what happens if it is a gray surface. For such a surface, the following additional relations hold.

$$\epsilon_\lambda' \neq f(\lambda) \qquad (3.26)$$

$$\epsilon'(T, \theta, \phi) = \epsilon_\lambda'(T, \theta, \phi) \qquad (3.27)$$

Therefore, the formula that we derived for hemispherical directional emissivity, as well as for directional total emissivity, when reduced to the case of a gray and diffuse body, respectively, reduce to the cases for which we are able to intuitively guess the values of ϵ_λ'. Therefore, the expressions we derived must be correct and these two expressions can be used to relate the fundamental emissivity ϵ_λ' to that quantity, which is the emissivity integrated once, either with respect to angle or with respect to wavelength. Once the triple integral with respect to θ, ϕ and λ is done, we get the hemispherical total emissivity ϵ.

3.4 Hemispherical Total Emissivity, $\epsilon(T)$

After we accomplish three integrations of ϵ_λ', once with respect to wavelength and other with respect to the angles (both azimuthal and zenith) we get the hemispherical total emissivity $\epsilon(T)$. Please note that the final emissivity will be a function of temperature. However, sometimes the dependence is weak and in which case, we assume that the emissivity is independent of temperature. The hemispherical, total emissivity denoted by $\epsilon(T)$ is given by the emissive power of a real surface at a given

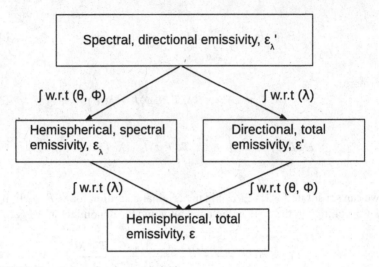

Fig. 3.2 Bird's eye view of various emissivities

temperature T divided by the emissive power of a black body at the same temperature.

$$\epsilon(T) = \frac{E(T)}{E_b(T)} \tag{3.28}$$

$$E(T) = \int_{\lambda=0}^{\infty} \int_{\phi=0}^{2\pi} \int_{\theta=0}^{\pi/2} \epsilon_\lambda'(\lambda, T, \theta, \phi) I_{b,\lambda}(\lambda, T) \cos\theta \sin\theta d\theta d\phi d\lambda \tag{3.29}$$

We know that

$$E_b(T) = \sigma T^4 \tag{3.30}$$

$$\therefore E(T) = \frac{1}{\sigma T^4} \int_{\lambda=0}^{\infty} \int_{\phi=0}^{2\pi} \int_{\theta=0}^{\pi/2} \epsilon_\lambda'(\lambda, T, \theta, \phi) I_{b,\lambda}(\lambda, T) \cos\theta \sin\theta d\theta d\phi d\lambda \tag{3.31}$$

Equation 3.31 represents a very important relationship in view of the fact that it tells us that given the spectral directional emissivity ϵ_λ', it is possible for us to do the 3 integrations and obtain ϵ. For a gray diffuse surface, ϵ_λ' is not a function of λ, θ or ϕ.

$$\epsilon_\lambda' \neq f(\lambda, \theta, \phi) \tag{3.32}$$

Therefore, ϵ_λ' can be taken out of all the 3 integrals in Eq. (3.31). Applying Stefan's law for the remaining terms within the integral, we get σT^4. Upon doing this and simplifying Eq. (3.31), we finally have

$$\epsilon(T) = \frac{1}{\sigma T^4} \epsilon_\lambda'(T)\sigma T^4 = \epsilon_\lambda'(T) \qquad (3.33)$$

Therefore, if we have the hemispherical directional emissivity for a gray, diffuse surface, it is its hemispherical total emissivity too. Figure 3.2 gives a bird's eye view of the various emissivities involved.

Example 3.1 The hemispherical spectral emissivity of tungsten is shown in Fig. 3.3 (this is an approximation of the actual variation and is sufficient enough to obtain reasonable estimates of $\epsilon(T)$). Consider a cylindrical tungsten filament that has a diameter of $D = 0.8$ mm and length $L = 25$ mm. The filament is enclosed in an evacuated bulb and is heated electrically till it reaches a steady state temperature of 3000 K.

(a). Determine the total hemispherical emissivity when the filament temperature is 3000 K.
(b). Determine the rate of cooling of the filament at the instant when the power is switched off? Tungsten properties are: $\rho = 19300$ kg/m³; $c_p = 132$ J/kgK. Assume the following: (1) surroundings are at 303 K (2) filament is spatially isothermal (3) neglect convection to the surroundings.

Fig. 3.3 Variation of spectral emissivity of tungsten with wavelength

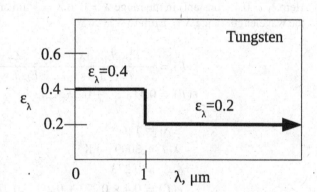

Solution:
The first part can be answered by directly using some formulae we have learned so far. The second part is a typical heat transfer problem, where using this emissivity and our knowledge of heat transfer, we will need to write the governing equation and obtain the initial rate of cooling.

(a). The first part involves the conversion of hemispherical spectral emissivity to hemispherical total emissivity.

$$\epsilon(T) = \frac{\int_{\lambda=0}^{\infty} \int_{\phi=0}^{2\pi} \int_{\theta=0}^{\pi/2} \epsilon_\lambda'(\lambda, T, \theta, \phi) I_{b,\lambda} \cos\theta \sin\theta \, d\theta \, d\phi \, d\lambda}{\sigma T^4} \qquad (3.34)$$

$$\epsilon_\lambda(\lambda, T) = \epsilon_\lambda = \frac{1}{\pi} \int\limits_{\phi=0}^{2\pi} \int\limits_{\theta=0}^{\pi/2} \epsilon_\lambda'(\lambda, T, \theta, \phi) \cos\theta \sin\theta d\theta d\phi \tag{3.35}$$

Equation (3.34) may be rewritten as follows:

$$\epsilon(T) = \frac{\int_{\lambda=0}^{\infty} I_{b,\lambda} d\lambda \int_{\phi=0}^{2\pi} \int_{\theta=0}^{\pi/2} \epsilon_\lambda'(\lambda, T, \theta, \phi) \cos\theta \sin\theta d\theta d\phi}{\sigma T^4} \tag{3.36}$$

Equation (3.36) may be simplified using Eq. (3.35) as

$$\epsilon(T) = \frac{\int_{\lambda=0}^{\infty} \epsilon_\lambda \pi I_{b,\lambda} d\lambda}{\sigma T^4} = \frac{\int_{\lambda=0}^{\infty} \epsilon_\lambda E_{b,\lambda} d\lambda}{E_b(T)} \tag{3.37}$$

Now we have to apply it to the tungsten function given in the question. The emissivity of tungsten can be written as

$$\epsilon(T) = \frac{\int_{\lambda=0}^{1} \epsilon_1 E_{b,\lambda} d\lambda}{E_b(T)} + \frac{\int_{\lambda=1}^{\infty} \epsilon_2 E_{b,\lambda} d\lambda}{E_b(T)} \tag{3.38}$$

Here $\epsilon_1 = 0.4$ (constant) in the range $\lambda = 0$ to $\lambda = 1$ μm and $\epsilon_2 = 0.2$ (constant) in the wavelength range $\lambda = 1$ μm to $\lambda = \infty$.

$$\epsilon(T) = \frac{\epsilon_1 \int_{\lambda=0}^{1} E_{b,\lambda} d\lambda}{E_b(T)} + \frac{\epsilon_1 \int_{\lambda=2}^{\infty} E_{b,\lambda} d\lambda}{E_b(T)} \tag{3.39}$$

$$\epsilon(T) = 0.4 F_{0 \to \lambda_1} + 0.2 F_{\lambda_1 \to \infty} \tag{3.40}$$

$$\lambda_1 = 1 \ \mu\text{m} \tag{3.41}$$

$$\lambda_1 T = 3000 \ \mu\text{mK} \tag{3.42}$$

$$F_{0 \to \lambda_1} = 0.273 \tag{3.43}$$

$$\epsilon(T) = 0.4 \times 0.273 + 0.2(1 - 0.273 \tag{3.44}$$

$$\epsilon(T) = 0.254 \tag{3.45}$$

In this example, we have thus far learnt how to get the hemispherical total emissivity, if the spectral emissivity is given for a non gray surface. If the ϵ_λ versus λ is completely jagged, we will have to convolve the Planck's function with the ϵ_λ and solve the problem numerically. If we have a band model like one specified in this problem, we can use the F-function chart and obtain $\epsilon(T)$ straightaway.

Here, if the temperature changes, even though ϵ_λ remains the same with respect to λ, if the same tungsten filament were to be at 2000 K, $\lambda_1 T = 2000$ μmK. Therefore the $\epsilon(T)$ will change and it can be seen that ϵ, in general, is a function of temperature. Here since 0–1 μm is such a small portion of the total electromagnetic spectrum, and for the remainder of the spectrum, ϵ_λ is only 0.2, we can actually calculate for various temperatures and see that for a range of temperatures, ϵ is more or less 0.2, because from 1 μm to ∞, ϵ_λ is 0.2.

(b) We have to calculate the initial cooling rate and for this, we have to get the energy equation first. For this, we will assume that the whole tungsten filament is at the same temperature and that convection heat losses are negligible.

$$mc_p\frac{dT}{dt} = -\epsilon\sigma A(T^4 - T_\infty^4) \tag{3.46}$$

When it starts cooling, the temperature of tungsten is at 3000 K while the temperature of the surroundings is 300 K. The other parameters in the above equation are known, and hence we can get the initial cooling rate. Even so this cooling rate will not remain a constant as the right side is a function of temperature. As the filament cools, the temperature will fall, and in turn, the cooling rate will fall and this is why it is called a non-linear function. Because the cooling rate is the rate of change of temperature and the rate of change of temperature is itself a function of the temperature, it is a non-linear function.

We calculate the volume and multiply it by the density to get the mass and then get the surface area (both the lateral surface area and the top and bottom areas). We then calculate the initial cooling rate, because the emissivity also changes with temperature. We can write a Matlab code to determine the cooling rate at various instants of time if this is desired. For the question at hand

$$m = \rho v \tag{3.47}$$

$$v = \frac{\pi}{4}d^2L = 1.256 \times 10^{-8} \text{ m}^3 \tag{3.48}$$

$$A = 2\pi rh + 2\pi r^2 \tag{3.49}$$

$$A = 6.383 \times 10^{-5} \text{ m}^2 \tag{3.50}$$

$$1.256 \times 10^{-8} \times 19300 \times 132\frac{dT}{dt} = -0.254 \times 6.383 \times 10^{-5} \times 5.67 \times 10^{-8}$$
$$\times [3000^4 - 300^4] \tag{3.51}$$

$$\frac{dT}{dt} = -3279 \text{ K/s} \tag{3.52}$$

We should not be carried away and think that in 1 s, the filament will lose 3279 K! This is just the initial cooling rate. In a few microseconds, because of the terrific cooling the temperature will go down, ϵ will go down and the rate reduces and reaches saner values. There are other properties like reflection, absorption and transmission and each of these may have a variation with respect to λ. We need to characterize all these and the energy equation may not be so simple. There may be combined conduction and convection. The energy equation may be such that we may have to solve the Navier-Stokes equations and the energy equation for a moving fluid or we may have to solve the Laplace equation or the Poisson equation for the solid wherein radiation is added so that it becomes a multimode heat transfer problem.

Example 3.2 The directional total emissivity of many non-metallic surfaces may be approximated, represented as $\epsilon_\theta = \epsilon_n \cos\theta$ where ϵ_n is the normal emissivity. Determine the ratio of the total, hemispherical emissivity and the emissivity at $\theta = 0$, frequently referred to as normal emissivity for one such material.

Solution:

$$\epsilon = \frac{\int_{\phi=0}^{2\pi} \int_{\theta=0}^{\pi/2} \epsilon_\theta I_b \cos\theta \sin\theta d\theta d\phi}{\int_{\phi=0}^{2\pi} \int_{\theta=0}^{\pi/2} I_b \cos\theta \sin\theta d\theta d\phi} \tag{3.53}$$

$$\epsilon = \frac{\epsilon_n 2\pi}{\pi} \int_{\theta=0}^{\pi/2} \cos^2\theta \sin\theta d\theta \tag{3.54}$$

$$\epsilon = -2\epsilon_n \int_{\theta=1}^{0} \cos^2\theta d(\cos\theta) \tag{3.55}$$

$$\epsilon = -2\epsilon_n \left[\frac{\cos^3\theta}{3}\right]_1^0 \tag{3.56}$$

$$\epsilon = \frac{2\epsilon_n}{3} \tag{3.57}$$

So given the directional total emissivity, we can get the total hemispherical emissivity.

Example 3.3 A zirconia-based ceramic is being considered for use as a candidate filament material for an incandescent bulb. It has a hemispherical spectral emissivity distribution as shown in Fig. 3.4.

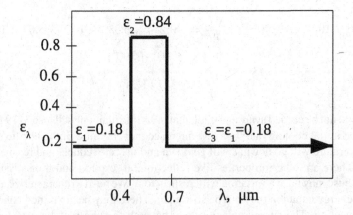

Fig. 3.4 Variation of spectral emissivity of zirconia with wavelength

1. Determine hemispherical, total emissivity of the zirconia filament at 2900 K?
2. Compare the power consumption of a zirconia and tungsten operating at 2900 K in an evacuated bulb ?
3. In so far as the production of visible radiation, which is more efficient? (adapted from Incropera et al. 2007).

Solution:

1. Hemispherical total emissivity ϵ
   $\lambda_1 = 0.4\ \mu m$, $\lambda_2 = 0.7\ \mu m$, $T = 2900$ K,
   $\lambda_1 T = 1160\ \mu mK$, $\lambda_2 T = 2030\ \mu mK$

$$\epsilon = \frac{\int_{\lambda=0}^{\infty} \epsilon_\lambda E_{b,\lambda} d\lambda}{\int_{\lambda=0}^{\infty} E_{b,\lambda} d\lambda} = \frac{\int_{\lambda=0}^{\lambda_1} \epsilon_1 E_{b,\lambda} d\lambda}{\int_{\lambda=0}^{\infty} E_{b,\lambda} d\lambda} + \frac{\int_{\lambda_1}^{\lambda_2} \epsilon_2 E_{b,\lambda} d\lambda}{\int_{\lambda=0}^{\infty} E_{b,\lambda} d\lambda}$$
$$+ \frac{\int_{\lambda_2}^{\lambda=\infty} \epsilon_2 E_{b,\lambda} d\lambda}{\int_{\lambda=0}^{\infty} E_{b,\lambda} d\lambda} \tag{3.58}$$

We have to use the F-function chart now.

$$\epsilon = \epsilon_1 F_{0\to\lambda_1} + \epsilon_2 [F_{0\to\lambda_2} - F_{0\to\lambda_1}] + \epsilon_3 [1 - \dot{F}_{0\to\lambda_2}] \tag{3.59}$$

From the chart

$$F_{0\to\lambda_1} = 1.77 \times 10^{-3} \tag{3.60}$$
$$F_{0\to\lambda_2} = 0.0711 \tag{3.61}$$

Now we will insert these values in the expression for emissivity.

$$\epsilon = (0.18 \times 1.77 \times 10^{-3}) + (0.84 \times (0.0711 - 1.77 \times 10^{-3}))$$
$$+ (0.18 \times (1 - 0.0711)) \tag{3.62}$$

$$\epsilon = 0.225 \tag{3.63}$$

Therefore, the hemispherical total emissivity of the zirconia filament is 0.225
2. Power consumption
$$Q = \epsilon \sigma A (T^4 - T_\infty^4) \tag{3.64}$$

From Problem 3.1, the total emissivity of tungsten at 2900 K,

$$\lambda_1 T = 1 \times 2900 = 2900\ \mu mK \tag{3.65}$$
$$F_{0-\lambda_1} = 0.250 \tag{3.66}$$
$$\epsilon(T) = 0.4(0.250) + 0.2(1 - 0.250) = 0.250 \tag{3.67}$$

Both the filaments are operating at 2900 K. Assuming that the ambient temperature is the same for both, Stefan Boltzmann's constant is the same for both the filaments, the areas are also the same. Therefore, the ratio of the power consumption is the ratio of their emissivities.

$$\frac{Q_{\text{zirconia}}}{Q_{\text{tung}}} = \frac{\epsilon_{\text{zirconia}}}{\epsilon_{\text{tung}}} = 0.225/0.250 = 0.904 \tag{3.68}$$

Therefore, the zirconia bulb consumes 90.4% the power consumed by the tungsten filament bulb, for the same temperature.

3. To determine which is more efficient in production of visible radiation, we need to find out the radiation emitted in the visible part of the spectrum. The radiation which is coming out of the bulb must be equal to σT^4 the black body fraction (corresponding to 0.4–0.7 μm band corresponding emissivity). This can be worked out for both the bulbs and since we are only looking at the ratios, we can keep σT^4 as such without substituting the numerical values. Radiation emitted in the visible part of the spectrum is thus

$$Q_{\text{visible}} = \epsilon_{\lambda} . (F_{\lambda_1 \to \lambda_2}) . \sigma T^4 \tag{3.69}$$

$$Q_{\text{visible,Zirconia}} = 0.84 \times (0.0711 - 1.77 \times 10^{-3}) \sigma T^4 \tag{3.70}$$

$$Q_{\text{visible,Tungsten}} = 0.4 \times (0.0711 - 1.77 \times 10^{-3}) \sigma T^4 \tag{3.71}$$

The value used for the emissivity is that which corresponds to the visible part of the spectrum for the material under consideration. This ϵ is 0.84 for the zirconia filament, while it is just 0.4 for the tungsten filament. Even though the tungsten filament has a higher total hemispherical emissivity as opposed to the zirconia filament, the zirconia filament, by virtue of its having a very high spectral emissivity of 0.84 exactly in the visible part of the spectrum, gives more visible radiation compared to the tungsten filament. Furthermore, its power consumption is also lower!.

So a zirconia filament is infinitely better than the tungsten filament. Of course, cost, availability and other properties have to be considered during production. Suffice it to say for now that from the point of view of radiation, the zirconia filament is better.

Example 3.4 Consider an arrangement as shown in Fig. 3.5 to detect radiation emitted by an elemental surface of area $A_1 = 6.25 \times 10^{-6}$ m^2 and temperature $T_1 = 1100$ K. The area of detector $A_2 = 4 \times 10^{-6}$ m^2. For the radiation emitted by A_1 at $\theta = 0$ (normal direction) at a distance of L = 0.4 m, the detector measures a radiant power of 1.5×10^{-6} W. Determine the directional total emissivity of A_1 at $\theta = 0$. Now the detector is moved horizontally to position b such that $\theta = 45°$. For this position, the detector measures a radiant power of 1.46×10^{-7} W. Can we comment on whether the surface 1 is a diffuse emitter?

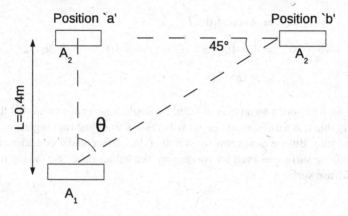

Fig. 3.5 Surface arrangements for Example 3.4

Solution:

$$E_b = \sigma T^4 = 5.67 \times 10^{-8} \times (1100)^4 \tag{3.72}$$

$$E_b = 83014.5 \ \text{W/m}^2 \tag{3.73}$$

$$I_b = \frac{\sigma T^4}{\pi} = 26432 \ \text{W/m}^2 sr \tag{3.74}$$

$$dw = \frac{A_2 cos\theta}{R^2} = \frac{4 \times 10^{-6}}{0.4^2} = 2.5 \times 10^{-5} sr \tag{3.75}$$

$$Q = 1.5 \times 10^{-6} \text{W} \tag{3.76}$$

$$Q = A\epsilon I_b dw \tag{3.77}$$

$$1.5 \times 10^{-6} = 6.25 \times 10^{-6} \times \epsilon \times 26432 \times 2.5 \times 10^{-5} \tag{3.78}$$

$$\epsilon = 0.36 \tag{3.79}$$

The normal emissivity = 0.36. This is one possible way by which we can measure emissivity. If we have a vacuum arrangement with a detector and we are able to eliminate the effects of conduction and convection, it is possible to get emissivity. The more difficult part is when the detector is moved horizontally. So cos(0) now changes to $cos\theta$. The distance, L also changes. Now because the detector is at an angle and the distance also changes, expectedly, the radiant power intercepted by A_2 has to go down.

$$R' = \frac{L}{cos\,\theta_2} = \sqrt{2}L = 0.566 \ \text{m} \tag{3.80}$$

$$d\omega' = \frac{A_2 \cos \theta_2}{R'^2} = 8.84 \times 10^{-6} sr \tag{3.81}$$

$$Q' = 1.46 \times 10^{-7} \ \text{W} \tag{3.82}$$

$$Q' = A_1 \cos\theta_1 d\omega' \, I_b \epsilon'_{\theta=45} \tag{3.83}$$

$$1.46 \times 10^{-7} = 4 \times 10^{-6} \times \frac{1}{\sqrt{2}} \times 8.84 \times 10^{-6} \times \epsilon' \times 26432 \tag{3.84}$$

$$\epsilon'_{\theta=45} = 0.142 \tag{3.85}$$

Suppose we had gotten an answer of 0.36, we could have conjectured that there is a possibility that it is a diffuse surface, as with results from just two angles, we cannot decide for sure. But we now know for sure that the surface under consideration here is not a diffuse surface as even for two angles, the values differ. So the emitter A_1 is NOT a diffuse surface.

3.5 Absorptivity, α

The next important property is absorptivity. Often times, we are also interested in the absorption and not just the emission. For example, if our application is a solar collector, where we want to intercept the solar radiation, we want to have a surface that absorbs a lot of radiation in the visible part of the spectrum. Once the surface starts absorbing, the temperature of the surface may go from the room temperature of 30 °C up to 80 °C or 90 °C. For such a surface from the Wien's displacement law, it is clear that the peak emission takes place in the infrared portion of the spectrum with a wavelength of about 7.5 μm. Now, if we have a surface that emits very poorly in the infrared part of the spectrum, but absorbs very well in the visible part of the spectrum, we have a good solar collector. By the same token, if we have a surface that absorbs very poorly in the visible part of the spectrum, but emits very well in the infrared part of the spectrum, it may be a good candidate for insulation.

First, we have to discuss about the story of radiation that is incident on a surface. **What can happen to this radiation?** Let us consider a surface on which radiation is incident (Fig.3.6). This radiation can be absorbed, reflected or transmitted. If we apply the first law of thermodynamics to this system, mathematically we can state that at steady state,

Incident radiation = absorbed radiation + reflected radiation + transmitted radiation.

$$Q_{inc} = Q_{abs} + Q_{ref} + Q_{trans} \tag{3.86}$$

Dividing by Q_{inc} throughout, we get

$$1 = \frac{Q_{abs}}{Q_{inc}} + \frac{Q_{ref}}{Q_{inc}} + \frac{Q_{trans}}{Q_{inc}} \tag{3.87}$$

On the right side, all 3 terms are dimensionless ratios, none of which individually can be greater than 1. They are also measures of the efficiency with which a surface absorbs, reflects or transmits.

Q_{abs}/Q_{inc} is called the absorptivity, denoted by α.

Q_{ref}/Q_{inc} is called the reflectivity, denoted by ρ.

Q_{trans}/Q_{inc} is called the transmissivity, denoted by τ.

Hence, Eq. (3.85) can be rewritten as

$$\alpha + \rho + \tau = 1 \qquad (3.89)$$

We can also write Eq. (3.89) for a particular wavelength in which case it becomes

$$\alpha_\lambda + \rho_\lambda + \tau_\lambda = 1 \qquad (3.90)$$

In Eq. 3.89, we are talking about hemispherical total quantities, as already, the integrations are done with respect to the angle and the wavelength. This is just to introduce the concepts. We will go through the definition of individual absorptivities and reflectivities a little later. For an opaque surface, $\tau=0$. Hence, Eq. 3.89 becomes

$$\alpha + \rho = 1 \qquad (3.91)$$
$$\alpha = 1 - \rho \qquad (3.92)$$

We have 4 quantities to deal with namely, emissivity, absorptivity, reflectivity and transmissivity. We "killed" one of them, namely transmissivity, for an opaque surface. We have 3 quantities left. We are now trying to see if further simplifications are possible. If emissivity is known, we can write α in terms of ρ. What then remains are only α and ϵ. Is there any relationship between these two? When the best minds were working on finding out the correct black body distribution, they were also looking at the properties of real surfaces and wondering if the emissivity and the absorptivity are related to each other. This was being studied as it makes a lot of things very convenient. Suppose we could establish a relationship between α and ϵ, and we were

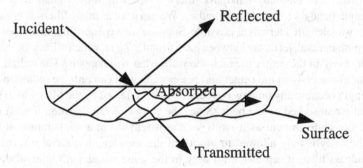

Fig. 3.6 Absorption, reflection and transmission processes associated with a semi-transparent medium

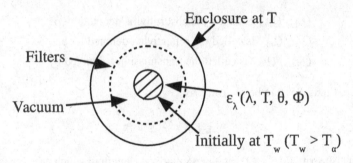

Fig. 3.7 Radiation in an enclosure with filters placed between a real body and the walls of the enclosure

able to measure the emissivity, then using this relationship, we can get α and also determine ρ using Eq. (3.92). Therefore, for an opaque surface, by just knowing the emissivity, we get all its properties that are required for carrying out a radiative transfer analysis.

Why do we think there should be a relationship between the emissivity ϵ and the absorptivity α? Can this relationship come from theory? Since the physical mechanisms of absorption and emission are different, we do not expect any logical relationship between emissivity ϵ and the absorptivity α to flow from theory, and hence this relationship must come from experiments. Fortunately, researchers have done experiments and have determined relationships, which we shall study in the next section.

Let us now carry out a thought experiment. Consider an evacuated enclosure at a temperature T as shown in Fig. 3.7. Now we have a body initially at a temperature T_w, where $T_w > T_\infty$. For a change, we do not have a black body, but we have a body whose spectral directional emissivity is given by $\epsilon'_\lambda(\lambda,T,\theta,\phi)$. What does the second law of thermodynamics tell us? If T_w is different from T_∞ and $T_w > T_\infty$, because there is vacuum, there will be no conduction or convection and only radiation will take place. Eventually, this body will also reach a temperature of T. We put some filters, which are basically bandpass filters which will allow radiation of narrow wavelength band $d\lambda$ to cross the boundary. We can have as many filters as possible in as many wavelength intervals as possible. Suppose we have one filter in one particular wavelength interval, let us say between 3.6 μm and 3.7 μm, which allows the radiation from the body in the center to reach the wall, what will happen? The radiation that goes out of the body in the center and reaches the wall can only be radiation in this wavelength because any other wavelength will be reflected by the filter which will be eventually reabsorbed by the body. Since the body is in equilibrium, it must absorb exactly the same amount as it emits so that it remains at a temperature of T. But since the body is only allowed to absorb in the wavelength interval $d\lambda$, thanks to the bandpass filter, it can also emit only in the same wavelength interval $d\lambda$. This $d\lambda$ is under our control and we can change it from say 3.6–3.7 μm or from 8.1 μm to 8.3 μm. Therefore, under these conditions, since the body is both emitting and absorbing in a particular wavelength interval, we can also choose the direction, by making the filter in such a way that it permits in only one direction. We are taking recourse to the second law of thermodynamics, which forbids the body from being at a temperature different from that of the surroundings because eventually equilibrium will be established.

$$dQ_{abs} = dQ_{emitted} \tag{3.93}$$

The right side can be evaluated using the formule discussed earlier. But to calculate dQ_{abs}, it is imperative for us to define spectral directional absorptivity.

3.6 Spectral Directional Absorptivity, α'_λ

The spectral directional absorptivity, α'_λ is defined as

$$\alpha'_\lambda(\lambda, T, \theta, \phi) = \frac{dQ_{abs}}{I_{\lambda,i} \cos\theta_i \sin\theta_i d\theta_i d\phi_i d\lambda dA} \tag{3.94}$$

Equation 3.93 clearly states that the spectral directional absorptivity is the radiation absorbed in a particular wavelength and direction to that incident at the same wavelength and direction. Needless to say, α'_λ too is a dimensionless quantity varying from 0 to 1.

Consider radiant balance from a surface of body similar to the one consider in Fig. 3.7. Let the surface under consideration be under equilibrium so that the net radiation heat transfer is zero. This can be written mathematically as follows

$$Q_{net} = Q_{outgoing} - Q_{incoming} = Q_{emitted} + Q_{reflected} - Q_{incident} \tag{3.95}$$

For this body, which is in equilibrium, $Q_{net} = 0$. If this Q_{net} were not equal to 0, the temperature of the body will increase or decrease with time. Therefore

$$Q_{emitted} + Q_{reflected} - Q_{incident} = 0 \tag{3.96}$$

If there is no transmission

$$Q_{incident} - Q_{reflected} = Q_{absorbed} \tag{3.97}$$
$$\therefore Q_{emitted} - Q_{absorbed} = 0 \tag{3.98}$$
$$\text{or} \tag{3.99}$$
$$Q_{emitted} = Q_{absorbed} \tag{3.100}$$

This was what we wrote in Eq. (3.93) where the quantities we introduced were differential ones. Apart from saying that the radiation will be only in the wavelength band $d\lambda$ about λ, we can also decide on the direction, by making the bandpass filter allow radiation in just one direction. Substituting the expressions in Eq. (3.100), we get

$$- \alpha(\lambda, T, \theta, \phi) I_{\lambda,i}(\lambda, \theta_i, \phi_i) \cos \theta_i \sin \theta_i d\theta_i d\phi_i d\lambda$$
$$= \epsilon'_\lambda(\lambda, T, \theta_i, \phi_i) I_{b,\lambda}(\lambda, T) \cos \theta_i \sin \theta_i d\theta_i d\phi_i d\lambda \qquad (3.101)$$

A spherical cavity with vacuum inside is equivalent to a black body. Therefore, the incident radiation, as far as the small object is concerned, is basically coming from the walls of the enclosure. There is perfect reflection among the various surfaces of the walls of the enclosure. Since the walls of the enclosure and the small body are at the same temperature, $I_{\lambda,i}$ corresponds to uniform or isotropic radiation from the walls of the enclosure which can be deemed to be radiation from a black body at a temperature T, which is the same as that of the small body within the enclosure. Therefore, we are allowed to change $I_{\lambda,i}$ to $I_{b,\lambda}$ on the left hand side of Eq. 3.101

$$I_{\lambda,i} = I_{b,\lambda} \qquad (3.102)$$

Upon doing this in Eq. (3.101) and cancelling the common terms on both sides, we get

$$\boxed{\alpha'_\lambda(\lambda, T, \theta, \phi) = \epsilon'_\lambda(\lambda, T, \theta, \phi)} \qquad (3.103)$$

Thus, the spectral directional absorptivity is equal to the spectral directional emissivity.

This is the **Kirchoff's law** which is always true and holds good without any constraints (Gustav Kirchoff (1824–1887, German physicist, published research papers on the thermal laws of radiation between 1959 and 1862).

The Kirchoff's law is general and is valid for any wavelength and any angle and is also applicable for situations where a surface need not be housed in an enclosure. This configuration was only used to prove this law. The law can be experimentally verified. What we have presented above is one kind of proof of the Kirchoff's law.

For a gray surface,

$$\epsilon'_\lambda, \alpha'_\lambda \neq f(\lambda) \qquad (3.104)$$
$$\therefore \epsilon'(T, \theta, \phi) = \alpha'(T, \theta, \phi) \qquad (3.105)$$

For a diffuse surface

$$\epsilon'_\lambda, \alpha'_\lambda \neq f(\theta, \phi) \qquad (3.106)$$
$$\therefore \epsilon'_\lambda(\lambda, T) = \alpha'_\lambda(\lambda, T) \qquad (3.107)$$

The angles θ_i and ϕ_i are very important because normally we can keep them as θ and ϕ. But when we are considering reflection, (here we are considering only absorption) there is a θ_i and ϕ_i and θ_r and ϕ_r. The radiation can be received in one direction, while the reflection can be in all directions or it can be specular (the radiation enters in one direction and goes out in one direction only). That is why in absorptivity, we use the subscript i for the angles. But in reflectivity, we will encounter both i and r subscripts for the angles, which makes a characterization of the reflectivity of surfaces more complicated.

For a gray, diffuse surface the spectral, directional and hemispherical, total quantities are one and the same. Hence, for such a surface, emissivity is equal to absorptivity with the understanding that here, we are referring to the hemispherical, total quantities. Equation (3.107) can be substantially simplified as

$$\epsilon = \alpha \tag{3.108}$$

The other key point to be notes is that Eq. (3.108) is not the Kirchoff's law. It is the post-processed version of the Kirchoff's law for a gray diffuse body. The Kirchoff's law in its basic form is more general as it states that the spectral directional emissivity is equal to the spectral directional absorptivity. For a gray, diffuse, opaque surface, $\epsilon = \alpha$ which is a very common engineering assumption and can be a "life saver" in many engineering applications. Suppose we want to determine the radiative heat transfer between the walls in a room, we start off with the knowledge that the walls of the room are opaque. Then if we make the assumption that these walls are made of gray, diffuse surfaces, the walls of the enclosure of this room can now be treated as a gray, diffuse, opaque enclosure which considerably simplifies the analysis. For such a wall

$$\tau = 0, \ \alpha = \epsilon \tag{3.109}$$
$$\rho = 1 - \alpha = 1 - \epsilon \tag{3.110}$$

From literature and from experiments, if we know what the emissivity of the surfaces is, then we can calculate the absorptivity and reflectivity and proceed with the detailed radiative heat transfer calculations. Before we proceed to the spectral hemispherical absorptivity, we will give a sneak peek into what all this will eventually lead to. The final goal of all this is that we should be in a position to calculate the net radiation heat transfer from any surface. This surface may be one that is isolated or maybe a part of several surfaces in an enclosure or furnace or combustion chamber and so on.

Net radiation heat transfer at an opaque surface:

We have incident radiation, reflected radiation, absorbed radiation and since the surface is taken as opaque, there is no transmitted radiation. Even so there is also an emission consequent upon the temperature of the surface being at a temperature greater than 0 K.

In view of the above, Net radiation heat transfer at the surface = Radiation that is going out—Radiation that is coming in.

Outgoing radiation = Reflected component + Emission component.
Incoming radiation = incident radiation.

∴ Net radiation heat transfer at the surface = Reflection + Emission − Incident.

If convection and conduction are ruled out, the net radiation heat transfer at the surface is given by

$$Q_{net} = Q_{emit} + Q_{ref} - Q_{inc} \qquad (3.111)$$

Example 3.5 The hemispherical spectral emissivity of a surface is as shown in Fig. 3.8. Draw the corresponding distributions for hemispherical spectral absorptivity α_λ and hemispherical spectral reflectivity ρ_λ.

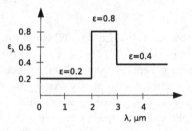

Fig. 3.8 Hemispherical spectral emissivity distribution

Solution:

The absorbtivity distribution will be the same as the emissivity distribution. The reflectivity is obtained as $1-\alpha$ (Fig. 3.9). However, calculating α from α_λ is not so straightforward. We need to know the variation of $I_{\lambda,i}$ with respect to λ. Or we should know from where that this $I_{\lambda,i}$ is coming from, as for example the sun. In this case, we are interested in calculating the solar absorptivity. For this case, the sun can be considered to be a black body at 5800 K. We can use the F-function chart and calculate absorptivity much in the same way as we calculate emissivity.

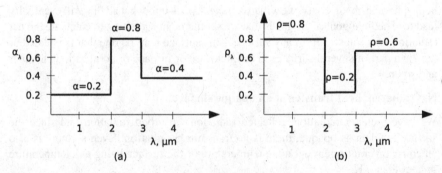

Fig. 3.9 Variation of spectral absorptivity and spectral reflectivity with wave length for Example 3.5

Fig. 3.10 Typical
representation for absorption
of radiation by a surface with
area dA

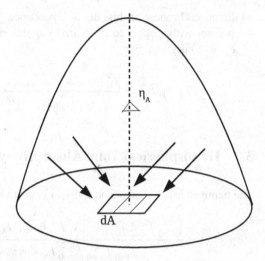

3.7 Hemispherical Spectral Absorptivity, $\alpha_\lambda(\lambda, T_A)$

If T_A is temperature of the absorbing surface, then the spectral, hemispherical absorptivity of the surface is defined as follows:

$$\alpha_\lambda(\lambda, T_A) = \frac{\int_{\phi_i=0}^{2\pi} \int_{\theta_i=0}^{\pi/2} \alpha'_\lambda(\lambda, T_A, \theta_i, \phi_i) I_{\lambda,i}(\lambda, \theta_i, \phi_i) \cos\theta_i \sin\theta_i d\theta_i d\phi_i}{\int_{\phi_i=0}^{2\pi} \int_{\theta_i=0}^{\pi/2} I_{\lambda,i}(\lambda, \theta_i, \phi_i) \cos\theta_i \sin\theta_i d\theta_i d\phi_i}$$

(3.112)

If we know the distribution of α'_λ as a function of θ and we also have information on $I_{\lambda,i}$ versus α'_λ, it is possible for us to multiply the two and integrate the product over the complete hemisphere to obtain the spectral hemispherical absorptivity.

The question before us is what is the goal of defining this quantity? Basically, we have an elemental surface of area dA. The unit vector is η_A. We are considering the hemispherical space above dA (see Fig. 3.10). Radiation from this hemispherical space is falling on this object dA. The ratio of the total radiation from the hemispherical space above at a particular wavelength interval which is absorbed by the body to the total radiation, coming from the hemispherical space, which is falling on this object in a particular wavelength interval is the hemispherical spectral absorptivity, again a dimensionless quantity varying between 0 and 1.

3.8 Directional Total Absorptivity, $\alpha'(T_A, \theta, \phi)$

The directional, total absorptivity is given by the ratio of the total radiation absorbed by a surface at all wavelengths in the (θ, ϕ) direction to that incident in the $(\theta,$

ϕ) direction. In view of this, the λ dependence goes because we will perform one integration with respect to λ to arrive at this quantity from α'_λ. Mathematically $\alpha'(T_A, \theta, \phi)$ is given by

$$\alpha'(T_A, \theta_i, \phi_i) = \frac{\int_{\lambda=0}^{\infty} \alpha'_A(\lambda, T_A, \theta_i, \phi_i) I_{\lambda,i}(\lambda, \theta_i, \phi_i) d\lambda d\omega}{\int_{\lambda=0}^{\infty} I_{\lambda,i}(\lambda, \theta_i, \phi_i) d\lambda d\omega} \tag{3.113}$$

3.9 Hemispherical Total Absorptivity, $\alpha(T_A)$

The hemispherical, total absorptivity, α is given by

$$\alpha(T_A) = \frac{\int_{\lambda=0}^{\infty} \int_{\phi_i=0}^{2\pi} \int_{\theta_i=0}^{\pi/2} \alpha'_\lambda(\lambda, T_A, \theta_i, \phi_i) I_{\lambda,i} \cos\theta_i \sin\theta_i d\theta_i d\phi_i d\lambda}{\int_{\lambda=0}^{\infty} \int_{\phi_i=0}^{2\pi} \int_{\theta_i=0}^{\pi/2} I_{\lambda,i} \cos\theta_i \sin\theta_i d\theta_i d\phi_i d\lambda} \tag{3.114}$$

Through the preceding development, we have reached the stage where if α'_λ is given to us, we have a mechanism to get α. If the distribution is known, we can integrate the numerator and denominator of Eq. 3.113 analytically or numerically and get α. This is of final engineering interest to us. With α and ϵ, we can work on the actual heat transfer problems which could involve modes other than radiative heat transfer. So, once we have information of α and ϵ, we have crossed two important hurdles in our pursuit of determining radiation heat transfer between surfaces. These are (i) the radiation laws and (ii) characterization of a surface that is not a black body.

The next two hurdles will be how to take care of the geometric orientation of the various objects and if many objects are involved, how do we take care of the overall formulation and what is the influence of one object on the other?

Getting back to Eq. 3.114, if the irradiation is from a black body that is at temperature T_S, which is the temperature of the sun, the denominator of Eq. (3.114) will become σT_S^4. Then α is no longer independent of the temperature of the surface from which the radiation is originating because T_S^4 is there in the denominator of this expression. Therefore, α becomes a function of T_A and T_S. This is valid even if the irradiation is from a diffuse gray object at a temperature T_S.

Hence, for the special case of irradiation from a black body at temperature T_S, the hemispherical total absorptivity, α, is given as

$$\alpha(T_A, T_S) = \frac{1}{\sigma T_S^4} \int \int \int \alpha'_\lambda I_{\lambda,i} \cos\theta \sin\theta d\theta d\phi d\lambda \tag{3.115}$$

The most profound and subtle change we have made on the left hand side of the equation is that we have made α a function of T_S too, outside of T_A.

Example 3.6 Consider an opaque surface with the spectral hemispherical absorptivity as shown in Fig. 3.11. The spectral distribution of incident radiation is also given in the figure.

(a) Determine the hemispherical total absorptivity of the surface?
(b) If this surface is diffuse and is at 1200 K, what is its total hemispherical emissivity?
(c) Determine the net radiation heat transfer from the surface?

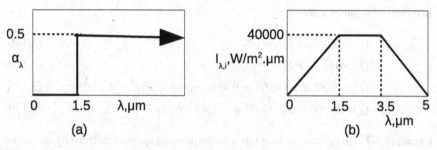

Fig. 3.11 Variation of spectral absorptivity and radiation intensity with wavelength for Problem 3.6

Solution:

(a) Hemispherical total absorptivity of the surface

$$\alpha = \frac{\int_{\lambda=0}^{\infty} \alpha_\lambda I_{\lambda_i} d\lambda \int \int \cos\theta \sin\theta d\theta d\phi}{\int_{\lambda=0}^{\infty} I_{\lambda_i} d\lambda \int \int \cos\theta \sin\theta d\theta d\phi} \qquad (3.116)$$

The total irradiance G in W/m^2 is given by

$$G = \int_{\lambda=0}^{\infty} I_{\lambda,i} d\lambda = \frac{1.5 \times 40000}{2} + 2 \times 40000 + \frac{1.5 \times 40000}{2} \qquad (3.117)$$

$$G = 140000 \ \text{W/m}^2 \qquad (3.118)$$

$$\alpha = \frac{0 \int_0^{1.5} I_{\lambda_i} d\lambda + 0.5 \int_{1.5}^{3.5} I_{\lambda,i} d\lambda + 0.5 \int_{3.5}^{5} I_{\lambda,i} d\lambda}{140000} \qquad (3.119)$$

$$\alpha = \frac{0.5 \times 80000 + 30000 \times 0.5}{140000} = 0.39 \qquad (3.120)$$

(b) Total hemispherical emissivity
For a diffuse surface,

$$\alpha_\lambda = \epsilon_\lambda \qquad (3.121)$$

$$\lambda_1 = 1.5 \,\mu m \tag{3.122}$$

$$\lambda_1 T = 1.5 \times 1200 = 1800 \,\mu mK \tag{3.123}$$

$$F_{0 \to \lambda_1 T} = 0.0393 \tag{3.124}$$

$$\epsilon(T) = \frac{\int_0^{1.5} \epsilon_\lambda E_{b,\lambda} d\lambda}{E_b(T)} + \frac{\int_{1.5}^\infty \epsilon_\lambda E_{b,\lambda} d\lambda}{E_b(T)} \tag{3.125}$$

$$= 0 + 0.5(1 - 0.0393) = 0.480$$

(c) Net radiation heat transfer

$$Q = G - \epsilon\sigma T^4 - \rho G \tag{3.126}$$

$$Q = \alpha G - \epsilon\sigma T^4 \tag{3.127}$$

$$Q = (0.39 \times 140000) - 0.480 \times 5.67 \times 10^{-8} \times 1200^4 \tag{3.128}$$

$$Q = 54600 - 56435.09 = -1.83 \,kW/m^2 \tag{3.129}$$

Example 3.7 An opaque surface has a hemispherical spectral reflectivity as shown in Fig. 3.12a. It is subjected to a spectral irradiation as shown in Fig. 3.12b.

1. Sketch the spectral hemispherical spectral absorptivity distribution.
2. Determine the total irradiation on the surface.
3. Determine the radiant flux that is absorbed by the surface.
4. Determine the total hemispherical absorptivity of the surface (Fig. 3.13).

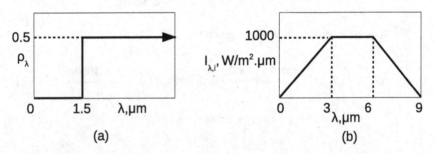

Fig. 3.12 a Hemispherical spectral reflectivity and **b** Spectral irradiance for Example 3.7

Solution:

1. For an opaque surface, $\tau_\lambda = 0$

$$\alpha_\lambda + \rho_\lambda = 1 \quad , \alpha_\lambda = 1 - \rho_\lambda \tag{3.130}$$

The spectral hemispherical spectral absorptivity distribution is shown in Fig. 3.13.

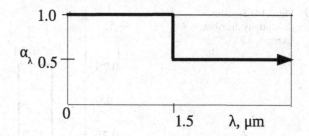

Fig. 3.13 Spectral
absorptivity distribution for
Example 3.7

2. The total incident radiation, I_{inci} in W/m² is the area under the curve.

$$I_{inci} = \left(\frac{1}{2} \times 3 \times 1000\right) + (3 \times 1000) + \left(\frac{1}{2} \times 3 \times 1000\right)$$

$$= 1500 + 3000 + 1500 = 6000 \, \text{W/m}^2 \quad (3.131)$$

3. Radiant flux absorbed by the surface

$$I_{abs} = \int_{\lambda=0}^{\infty} \alpha_\lambda I_{\lambda,i} d\lambda \quad (3.132)$$

$$I_{abs} = \frac{(1 \times 1500)}{2} + 0.5 \times 3000 + 0.5 \times 3000 \quad (3.133)$$

$$I_{abs} = 3000 \, \text{W/m}^2 \quad (3.134)$$

4. Total, hemispherical absorptivity of the surfce

$$\alpha = \frac{I_{abs}}{I_{inc}} = \frac{3000}{6000} = 0.5 \quad (3.135)$$

Example 3.8 A spatially isothermal surface is maintained at a temperature of 120 °C. Solar radiation with $I_{incident}=1050 \, \text{W/m}^2$, is incident at the top of the surface. The surface has an area of 2.55 m². The surface is opaque and diffuse and its spectral hemispherical absorptivity is given in Fig. 3.14. Determine

(a) the absorbed irradiation
(b) the emissive power
(c) net radiation heat transfer from the surface.

Fig. 3.14 Spectral absorptivity distribution for Example 3.8

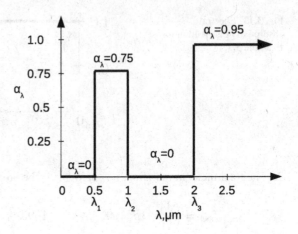

Solution:

(a) Absorbed irradiation

$$\alpha = \frac{\int_{\lambda=0}^{\infty} \alpha_\lambda E_{b,\lambda} \mathrm{d}\lambda}{\int_{\lambda=0}^{\infty} E_{b,\lambda}} \qquad (3.136)$$

Since nothing is specified about the variation of α_λ or ϵ_λ with respect to temperature, we assume that the same calculation holds good for all the temperatures. From F-function charts

$$\lambda_1 T = 0.5 \times 5800 = 2900 \,\mu\mathrm{mK} \qquad (3.137)$$
$$\lambda_2 T = 5800 \,\mu\mathrm{mK} \qquad (3.138)$$
$$\lambda_3 T = 11600 \,\mu\mathrm{mK} \qquad (3.139)$$
$$F_{0 \to \lambda_1} = 0.250 \qquad (3.140)$$
$$F_{\lambda_1 \to \lambda_2} = [0.720 - 0.250] = 0.47 \qquad (3.141)$$
$$F_{\lambda_3 \to \infty} = 0.06 \qquad (3.142)$$
$$\alpha = 0 + (0.47 \times 0.75) + (0.95 \times 0.06) \qquad (3.143)$$
$$\alpha = 0.41 \qquad (3.144)$$

If we remember, nearly 37% of the radiation was concentrated in the visible region of the solar spectrum. The wavelength band was 0.4–0.7 μm. Here, we have a range of 0.5–1.0 μm. So if 0.4–0.7 μm were to be 37%, we expect 0.5–1.0 μm to be about 50%. Out of this 50%, its efficiency is 0.75. So we will get a value of about 0.38 and anyway beyond 2 μm, the emission fraction is only 6%. Even though the surface has a terrific α of 0.95 for $\lambda > 2$ μm, there is not much incoming radiation in that part of the spectrum. So, instead of just mechanically and routinely calculating with the F-function chart, with some insight, just by looking at the distribution, we will be able to estimate α for a given temperature. Beyond 2 μm, α_λ can be anything whose does not really affect our calculations of the hemispherical, total absorptivity in this problem.

$$G_{abs} = 0.41 \times 1050 = 429.97 \ \text{W/m}^2 \tag{3.145}$$

Absorbed irradiation $= 429.97 \times 2.55 = 1096$ W.

(b) Emissive power

$$T = 393K, \ \lambda_1 T = 196.5 \ \mu\text{mK}, \ F_{0\to\lambda_1} = 0 \tag{3.146}$$
$$\lambda_2 T = 393 \ \mu\text{mK}, \ F_{\lambda_1\to\lambda_2} = 0 \tag{3.147}$$
$$\lambda_3 T = 786 \ \mu\text{mK}, \ F_{\lambda_3\to\infty} = 1.64 \times 10^{-5} \tag{3.148}$$
$$\epsilon = 0.95 \tag{3.149}$$

That is the way it should be because the surface temperature is 393 K and if we use the Wien's displacement law, $\lambda_{max} T = 2898 \ \mu\text{mK}$. Hence, here T = 393 K. λ_{max}, in this case, would be 7 μm. In this problem, we have said that after 2μm, there is no change, and hence even without using the F-function chart, we can say that all the distribution below 2 μm given here is irrelevant and (its effect is to the extent of 1.64×10^{-5}).

$$\text{Radiation emitted} = \epsilon \sigma T^4 A$$
$$= 0.945 \times 5.67 \times 10^{-8} \times 2.55 \times 393^4$$
$$= 3276.49 \ \text{W} \tag{3.150}$$

(c) Net radiation heat transfer from the surface

$$\text{Net radiation} = \text{Emitted} + (\text{Reflected} - \text{Incident})$$
$$= \text{Emitted} - \text{Absorbed}$$
$$= 3276.49 - (0.41 \times 1050 \times 2.55)$$
$$= 2178.71 W \tag{3.151}$$

Example 3.9 Solar flux of 950 W/m^2 is incident on the top surface of a plate whose solar absorptivity is 0.9 and emissivity is 0.1. The air and surroundings are at 27 °C and the convection heat transfer coefficient between the plate and air is 9 W/m^{2}K. Assuming that the bottom side of the plate is perfectly insulated, determine the steady state temperature of the plate. (Refer Fig. 3.15)

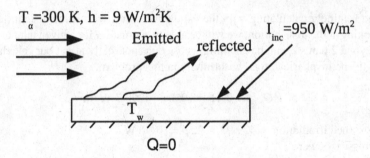

Fig. 3.15 Various heat transfer processes associated with Example 3.9

Solution:

The energy equation for this scenario will be

Net emission + Reflection + Convection = Incident (or)

Net emission + Convection = Net absorbed

$$\epsilon\sigma(T^4 - T_\infty^4) + \rho G_{inc} + h(T - T_\infty) = G_{inc} \qquad (3.152)$$

$$0.1 \times 5.67 \times 10^{-8}(T^4 - 300^4) + 9(T - 300) = 0.9 \times 950 \qquad (3.153)$$

$$T = \frac{[0.9 \times 950 - 0.1 \times 5.67 \times 10^{-8}(T^4 - T_\infty^4)]}{9} + 300 \qquad (3.154)$$

This is a non-linear equation which has to be solved by iterations. It is called a transcendental equation because it has a T^4 term. Equation (3.154) can be solved by the Newton-Raphson method or the successive substitution method or any other root extraction method.

The algorithm for successive substitution can be written as

$$T_{i+1} = \frac{[0.9 \times 950 - 0.1 \times 5.67 \times 10^{-8}(T_i^4 - 300^4)]}{9} + 300 \qquad (3.155)$$

We will start with 320 K and do iterations using the method of successive substitution.

After three iterations the temperature of the surface T is obtained as 386 K. Details of the iteration process are provided in Table 3.1.

Table 3.1 Successive substitution method for Example 3.9

Iteration no.	T_i	T_{i+1}
1	320	393.5
2	393.5	385
3	385	386.26
4	386.26	386.08

3.10 Reflectivity, ρ

We have already worked out problems based on this, but reflectivity is much more difficult and involved than we think. Radiation coming on to a surface can come from one particular direction or from the hemisphere above the surface. By the same token, we can look at the reflected radiation as going out in a particular direction or as going out in the hemispherical space above the surface; so there are 4 possibilities here.

(1) If incident radiation is from a particular direction, reflected radiation can go out in any direction.
(2) The incoming radiation can be from a particular direction while the radiation going out is into the hemispherical space.
(3) The incoming radiation can come from the hemispherical space and the outgoing radiation can be in any direction.
(4) Finally, radiation can come from the hemispherical space and also go out into the hemispherical space.

To all this, we add the spectral dependence. Consider Fig. 3.16 and let the incident radiation $I_{\lambda,i}$ be a function of λ, θ_i and ϕ_i. The outgoing radiation $dI_{\lambda,r}$ is a function of λ, the temperature of the surface T_A, θ_i, ϕ_i, θ_r and ϕ_r. $I_{\lambda,i}$ is incident from a particular angle θ_i, ϕ_i. The reflection can take place in any direction but we are looking at the reflected radiation in an angle θ_r, ϕ_r. As opposed to absorptivity, there is a significant

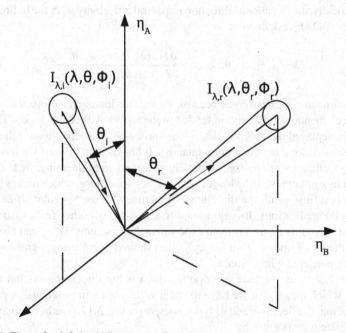

Fig. 3.16 Figure for defining bidirectional reflectivity

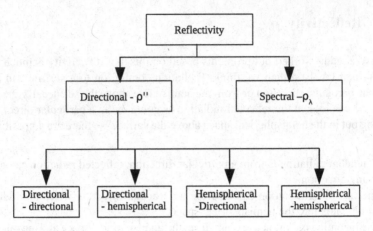

Fig. 3.17 Pictorial representation of reflectivity and its types

departure here because the reflected radiation is qualified by two additional variables θ_r and ϕ_r. The solid angle around (θ_i, ϕ_i) is small. Therefore, we expect the reflected component to be small as opposed to the incident. Therefore, we denote the reflected radiation by $dI_{\lambda,r}$.

The incoming radiation is from an angle (θ_i, ϕ_i), and outgoing is from (θ_r, ϕ_r). We are still discussing spectral quantities and hence the reflectivity under consideration should be a directional-directional spectral quantity. Therefore, it should be ρ_λ''. Mathematically, the directional-directional spectral reflectivity ρ_λ'' or the bidirectional reflectivity (BDRF) is defined as

$$\rho''(\lambda, T_A, \theta_i, \phi_i, \theta_r, \phi_r) = \frac{dI_{\lambda,r}(\lambda, T_A, \theta_i, \phi_i, \theta_r, \phi_r)}{I_{\lambda,i}(\lambda_i, \theta_i, \phi_i) \cos\theta_i d\omega_i} \qquad (3.156)$$

In the denominator, we have $\cos\theta_i$ because we are considering the projected area and $d\omega_i$ is the elemental solid angle subtended in the reflected direction θ_r, ϕ_r. This $d\omega_i$ can also be replaced by $\sin\theta_i \, d\theta_i \, d\phi_i$. Please note that the bidirectional reflectivity, by definition, is not a dimensionless quantity. It has the unit sr^{-1}. If we have good laboratory facilities, it can be experimentally measured. In engineering, we are mostly interested in ρ or at most, ρ_λ. Most of our surfaces are diffuse, which means that the reflection is diffuse in all the directions. Sometimes we use specular surfaces, for some special applications. It is important to know that all other reflectivities have their origin from this ρ_λ''. Apart from experimental measurements, ρ_λ'' can also come from the theory of optics. In summary ρ_λ'' has limited engineering significance but has a lot of conceptual importance.

For the first time, we have a property which is not dimensionless but has the unit sr^{-1}, which means that we have to use it with caution in problems. A pictorial representation of reflectivity and its types is shown in Fig. 3.17. The bi-hemispherical spectral reflectivity is given by

$$\rho_\lambda(\lambda, T_A) = \frac{\int_{\phi_r=0}^{2\pi} \int_{\theta_r=0}^{\pi/2} I_{\lambda,r}(\lambda, T_A, \theta_i, \phi_i, \theta_r, \phi_r) \cos\theta_r \sin\theta_r d\theta_r d\phi_r}{\int_{\phi_i=0}^{2\pi} \int_{\theta_i=0}^{\pi/2} I_{\lambda,i}(\lambda, T_A, \theta_i, \phi_i, \theta_r, \phi_r) \cos\theta_i \sin\theta_i d\theta_i d\phi_i} \quad (3.157)$$

The $I_{\lambda,r}$ in the numerator can be written as $I_{\lambda,i}$ multiplied by ρ'_λ where ρ'_λ can be directional to hemispherical or hemispherical to directional. This ρ'_λ can be connected to ρ''_λ which is directional-directional reflectivity. In short, if we know the bidirectional reflectivity ρ''_λ, it is possible for us to accomplish the integration and calculate the numerator. If we know the directional distribution of the incident radiation $I_{\lambda,i}$, we can calculate the denominator. Hence, we can straightaway get ρ_λ. If α_λ information is not available to us but if ρ''_λ is given, we can calculate ρ_λ and if it is an opaque surface, $1-\rho_\lambda$ can be taken to be α_λ. Things get complicated only when $\tau \neq 0$.

The bi-hemispherical total reflectivity ρ is given by

$$\rho = \frac{\int_{\lambda=0}^{\infty} \rho_\lambda I_{\lambda,i} d\lambda}{\int_{\lambda=0}^{\infty} I_{\lambda,i} d\lambda} \quad (3.158)$$

If the $I_{\lambda,i}$ distribution is not given, but we say that the incident radiation is from a black body at 5800 K, then we can convert it to $E_{b,\lambda}$ and use the F-function chart, use the α_λ and take $\rho_\lambda = 1-\alpha_\lambda$ and then calculate ρ. If in a problem, we have both ρ_λ and τ_λ, we can use $\alpha+\rho+\tau = 1$ and hence calculate α. So seamlessly, we should be able to go from one property to the other.

3.11 Transmissivity, τ

If we consider a material like glass, radiation incident on it can either be reflected, absorbed or transmitted. For a transparent or semi-transparent medium (i.e. a medium that is not opaque), there is a possibility that the radiation will penetrate the medium and come out of it. Therefore, an additional property enters, which is the transmissivity.

Transmissivity (τ) is also a dimensionless property and can have a value between 0 and 1. For an opaque surface, $\tau = 0$. If some surface has 100% transmittance, then $\tau = 1$. Just like emissivity, absorptivity and reflectivity, this can also vary with λ, θ and can lead to complexities. However, we often deal with thin media or a thin layer of glass or the atmosphere. The atmosphere is "thin" because compared to the radius of the Earth which is 6378 km, the thickness of the atmosphere is only 80 or 90 km. The height of the atmosphere divided by the radius is so small that the atmosphere can also be said be to a thin layer. If the atmosphere is considered to be thin, we can treat the resulting radiation problem as one dimensional along the height direction. We can consider all properties to vary only with the height or z-axis. That makes matters simple. If everything varies only with z, then the variation of transmissivity with angle does not have to be considered. Therefore, the concept of directional transmissivity becomes redundant for a one-dimensional medium or a plane parallel medium. So, wherever possible, we make a plane parallel assumption. If t is the thickness of the medium and L is its length, then if $L/t >> 1$, the medium is said to be a **plane parallel**. Properties vary only across the thickness and the term directional transmissivity is superfluous.

3.12   Spectral Transmissivity $\tau_\lambda(\lambda, t)$

A medium that can absorb, scatter and transmit is called a participating medium. If t is the thickness of the participating medium, then the spectral transmissivity

$$\tau_\lambda(\lambda, t) = \frac{I_{\lambda,tr}(\lambda)}{I_{\lambda,i}(\lambda)} \tag{3.159}$$

The basic difference between a participating medium and the surfaces so far considered is that reflection, absorption, etc. were all happening in the first few micrometers of the surface and being surface phenomena, were called radiative surface properties, However, in other situations, as for example, in the atmosphere, radiation penetrates deep inside, agitates all the molecules within it. In view of this, the atmosphere may absorb, scatter or reflect the radiation or the atmosphere itself may emit volumetrically. This is different from the emission of a surface. In the case of the atmosphere, the whole gas volume emits. It may scatter differently in different directions, which is called anisotropic scattering. The governing equation is called the **Radiative Transfer equation or RT equation**.

Getting back to transmissivity, the total transmissivity τ is given by

$$\tau(t) = \frac{\int_{\lambda=0}^{\infty} \tau_\lambda I_{\lambda,i} d\lambda}{\int_{\lambda=0}^{\infty} I_{\lambda,i} d\lambda} \tag{3.160}$$

Consider the case of solar irradiation on a semi-transparent surface. If we have the spectral distribution of ρ and α but we do not give any further information on τ, still τ can be calculated as follows:

$$\alpha + \rho + \tau = 0 \tag{3.161}$$

From the graphical distribution given for ρ and α, using the relation $\tau = 1-(\rho+\alpha)$, we can reconstruct the distribution for τ. For $I_{\lambda,i}$, we will take the $E_{b,\lambda}$ corresponding to the sun's temperature. Using the F-function chart, we can finally calculate τ.

In participating media, there can be emission, absorption and also scattering. Scattering is basically reflection, but is from the volume. There can be, in scattering, which is reflection on to the surface, and out scattering, which is reflection away from the surface and the net will be the difference between the out scattering and the in scattering. This scattering could be a function of wavelength and also a function of the angle. If it is not, then the medium is termed as isotropic. Scattering also depends on the size of the particle, like in the case of the atmosphere. Scattering by a dust particle will be different from the scattering by a water molecule, which will be different from the scattering by an ice particle. When the ice particle is oblique and non spherical, the scattering phenomena can get very involved.

Example 3.10 The spectral absorptivity α_λ and the spectral reflectivity ρ_λ for a diffuse surface are given in the Fig. 3.18.

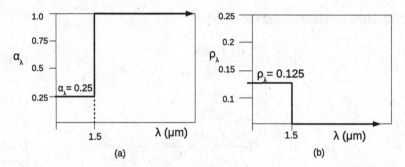

Fig. 3.18 Variation of spectral absorptivity and spectral reflectivity with wavelength for Problem 3.10

a. Sketch the spectral transmissivity distribution.
b. If solar radiation $G = 800 \text{ W/m}^2$ and spectral distribution corresponding to a black body at $6000K$ is incident on the material, determine the fractions of the irradiation that are absorbed, reflected and transmitted by the material.

Solution:

(a) Spectral transmissivity distribution

$$\alpha_\lambda + \rho_\lambda + \tau_\lambda = 1 \tag{3.162}$$
$$\tau_\lambda = 1 - (\alpha_\lambda + \rho_\lambda) \tag{3.163}$$
$$\tau_\lambda = 1 - (0.25 + 0.125) = 0.625 \tag{3.164}$$

The spectral transmissivity distribution is shown in Fig. 3.19.

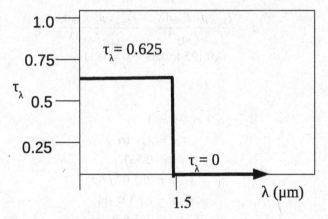

Fig. 3.19 Variation of spectral transmissivity with wavelength (Problem 3.10)

(b) Absorption, reflection and transmission

$$\alpha = \frac{\int_{\lambda=0}^{\infty} \alpha_\lambda . I_{\lambda,i} d\lambda}{\int_{\lambda=0}^{\infty} I_{\lambda,i} d\lambda} \tag{3.165}$$

$$\alpha = \frac{\int_{\lambda=0}^{\infty} \alpha_\lambda . E_{b\lambda} d\lambda}{\int_{\lambda=0}^{\infty} E_{b\lambda} d\lambda} \tag{3.166}$$

$$\alpha = \frac{\int_{\lambda=0}^{1.5} \alpha_\lambda . E_{b\lambda} d\lambda + \int_{\lambda=1.5}^{\infty} \alpha_\lambda . E_{b\lambda} d\lambda}{\int_{\lambda=0}^{\infty} E_{b\lambda} d\lambda} \tag{3.167}$$

$$\alpha = \frac{0.25 \int_{\lambda=0}^{1.5} E_{b\lambda} d\lambda + 1 \int_{\lambda=1.5}^{\infty} E_{b\lambda} d\lambda}{\int_{\lambda=0}^{\infty} E_{b\lambda} d\lambda} \tag{3.168}$$

$$\alpha = \frac{0.25 \int_{\lambda=0}^{1.5} E_{b\lambda} d\lambda}{\int_{\lambda=0}^{\infty} E_{b\lambda} d\lambda} + \frac{1 \int_{\lambda=1.5}^{\infty} E_{b\lambda} d\lambda}{\int_{\lambda=0}^{\infty} E_{b\lambda} d\lambda} \tag{3.169}$$

$T_s = 6000$ K (Given)

$$\lambda T = 1.5 \times 6000 = 9000 \, \mu mK \tag{3.170}$$
$$F_{0-\lambda T} = 0.89 \tag{3.171}$$
$$\alpha = 0.25 \times 0.89 + 1[1 - 0.89] = 0.332 \tag{3.172}$$

Solar absorptivity $= 0.332$

$$\rho = \frac{\int_{\lambda=0}^{\infty} \rho_\lambda . I_{\lambda,i} d\lambda}{\int_{\lambda=0}^{\infty} I_{\lambda,i} d\lambda} = \frac{\int_{\lambda=0}^{\infty} \rho_\lambda . E_{b\lambda} d\lambda}{\int_{\lambda=0}^{\infty} E_{b\lambda} d\lambda} \tag{3.173}$$

$$\rho = \frac{\int_{\lambda=0}^{1.5} \rho_\lambda . E_{b\lambda} d\lambda + \int_{\lambda=1.5}^{\infty} \rho_\lambda . E_{b\lambda} d\lambda}{\int_{\lambda=0}^{\infty} E_{b\lambda} d\lambda} \tag{3.174}$$

$$\rho = \frac{\int_{\lambda=0}^{1.5} \rho_\lambda . E_{b\lambda} d\lambda}{\int_{\lambda=0}^{\infty} E_{b\lambda} d\lambda} + \frac{\int_{\lambda=1.5}^{\infty} \rho_\lambda . E_{b\lambda} d\lambda}{\int_{\lambda=0}^{\infty} E_{b\lambda} d\lambda} \tag{3.175}$$

$$\rho = 0.125 \times 0.89 + 0 = 0.111 \tag{3.176}$$

Now

$$\alpha + \rho + \tau = 1 \tag{3.177}$$
$$\tau = 1 - (\alpha + \rho) \tag{3.178}$$
$$\tau = 0.557 \tag{3.179}$$
$$G_{abs} = 265.6 \text{ W/m}^2 \tag{3.180}$$
$$G_{ref} = 88.8 \text{ W/m}^2 \tag{3.181}$$
$$G_{trans} = 445.6 \text{ W/m}^2 \tag{3.182}$$

Example 3.11 The spectral transmissivities of plain and tinted glass vary non-linearly with wavelength. However, as a first cut approximation the following distribution may be assumed.

Plain glass: $\tau_\lambda = 0.9, 0.3 \leq \lambda \leq 2.5$ μm

Tinted glass: $\tau_\lambda = 0.9, 0.5 \leq \lambda \leq 1.5$ μm

Elsewhere, the spectral transmissivity is 0 for both the glasses.

  a. Compare the solar energy that is transmitted through the two glasses.

  b. If solar radiation is incident on the two glasses, compare the visible radiation that is transmitted by the two glasses.

  c. Comment on whether tinting the glass helps or hurts. (τ_λ values taken from Incropera et al. 2007)

Solution:

(a) Plain glass:

$$\tau = \frac{\int_{\lambda=0}^{\infty} \tau_\lambda . I_{\lambda,i} d\lambda}{\int_{\lambda=0}^{\infty} I_{\lambda,i} d\lambda} = \frac{\int_{\lambda=0}^{\infty} \tau_\lambda . E_{b\lambda} d\lambda}{\int_{\lambda=0}^{\infty} E_{b\lambda} d\lambda} \tag{3.183}$$

Let $T_s = 6000$ K

$$\lambda_1 \times T_s = 0.3 \times 5800 = 1740 \text{ μmK} \tag{3.184}$$

$$\lambda_2 \times T_s = 2.5 \times 5800 = 14500 \text{ μmK} \tag{3.185}$$

$$F_{0-\lambda_1 T_s} = 0.0326 \tag{3.186}$$

$$F_{0-\lambda_2 T_s} = 0.96643 \tag{3.187}$$

$$F_{\lambda_1-\lambda_2} = 0.968 - 0.039 = 0.929 \tag{3.188}$$

$$\tau = 0 \times 0.0326 + 0.9 \times (0.96643 - 0.0326) = 0.84 \tag{3.189}$$

So, for plain glass, 84% of the incident energy is transmitted. For tinted glass:

$$\lambda_1 = 0.5 \text{ μm}; \lambda_1 \times T_s = 2900 \text{ μmK} \tag{3.190}$$

$$\lambda_2 = 1.5 \text{ μm}; \lambda_2 \times T_s = 8700 \text{ μmK} \tag{3.191}$$

$$F_{0-\lambda_1 T_s} = 0.250 \tag{3.192}$$

$$F_{0-\lambda_2 T_s} = 0.8806 \tag{3.193}$$

$$\tau = 0 \times 0.253 + 0.9 \times (0.8806 - 0.250) = 0.567 \tag{3.194}$$

For tinted glass, only 56.7% of the incident energy passes through. So if the incident radiation is 1000 W/m², plain glass will allow 840 W/m² to pass through while the tinted glass allows 567 W/m² to pass through. So compared to plain glass, tinted glass stops 273 W/m² more which means the reduction is 27.3% in this case. This is important as the current trend everywhere is the use of glass and steel structures. Therefore, to reduce the air conditioning load, some solutions like this need to be used.

(b) Performance in the visible part of the spectrum

Plain glass: Within the visible band, $\tau_\lambda = 0.9$

$$\lambda_1 \times T_s = 0.4 \times 5800 = 2320 \ \mu mK \tag{3.195}$$
$$\lambda_2 \times T_s = 0.7 \times 5800 = 4060 \ \mu mK \tag{3.196}$$
$$F_{0-\lambda_1 \times T_s} = 0.1239 \tag{3.197}$$
$$F_{0-\lambda_2 \times T_s} = 0.4913 \tag{3.198}$$
$$\therefore \tau = 0 \times 0.1239 + 0.9 \times (0.4913 - 0.1239) = 0.33 \tag{3.199}$$

Tinted glass:

$$\lambda_1 \times T_s = 0.5 \times 5800 = 2900 \ \mu mK \tag{3.200}$$
$$\lambda_2 \times T_s = 0.7 \times 5800 = 4060 \ \mu mK \tag{3.201}$$
$$F_{0-\lambda_1 T_s} = 0.250 \tag{3.202}$$
$$F_{0-\lambda_2 T_s} = 0.4913 \tag{3.203}$$
$$\tau = 0 \times 0.250 + 0.9 \times (0.4913 - 0.250) = 0.217 \tag{3.204}$$

(c) Tinted glass cuts out the visible part by 12%. So, tinting the glass certainly helps in reducing the solar load.

3.13 Optical Pyrometry

We can devise instruments to measure the temperature of a surface based on its emission characteristics. This becomes the basis of an optical pyrometer. Suppose the surface is in the background, we can devise a system of lenses and focus on the radiation coming from the background. Consider an arrangement where a filament is placed ahead of the surface. The background is at some temperature. The temperature of the filament can be controlled by connecting it to a power source, as shown in Fig. 3.20. We keep increasing the filament's temperature such that its characteristics also change. We can change the settings such that at a particular point, this filament has the same temperature as the background. Since the two will merge, the filament will disappear from our field of view. At that point of time, the temperature of the filament is exactly the same as the temperature of the background and hence this is a way of inferring the temperature remotely. This is called pyrometry. Similarly, based on radiation and Planck's law too, equipment can be devised to measure the temperature remotely. Based on infrared radiation emitted from the objects, we can infer the temperature. So, first, calibration is done where each colour is calibrated against a temperature and from the colour the temperature is deduced. So if we look at all equipment we come across in heat transfer, their design is based on some particular law. They are adequately calibrated and benchmarked to make readings obtained from them trustworthy.

Fig. 3.20 Schematic of a
vanishing filament optical
pyrometer

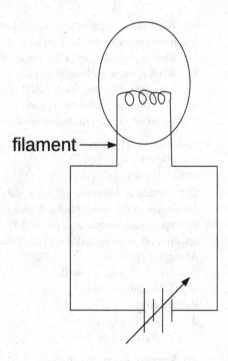

filament ——►

Problems

3.1 The hemispherical, spectral emissivity, ϵ_λ for a metal at 1000 K is approximately
given by
$0 \leq \lambda \leq 2\,\mu m$, $\epsilon_\lambda = 0.6$
$2 \leq \lambda \leq 4\,\mu m$, $\epsilon_\lambda = 0.35$
$\lambda \geq 4\,\mu m$, $\epsilon_\lambda = 0.15$
The ϵ_λ values do not change significantly with temperature and the metal surface
may be assumed to be diffuse.

(a) What is the hemispherical, total emissivity of the surface at 1000 K?
(b) If radiation is incident from a black body at 1400 K, what is the value of the
hemispherical total absorptivity for the incident radiation?
(c) If the irradiation due to the black body at 1400 K is 9000 W/m², what is the
net radiation heat transfer from the surface?

3.2 The hemispherical spectral emissivity, ϵ_λ for a metal at 1200 K is approximately
given by
$0 \leq \lambda \leq 2.5\,\mu m$, $\epsilon_\lambda = 0.75$
$2.5 \leq \lambda \leq 5\,\mu m$, $\epsilon_\lambda = 0.55$
$5 \leq \lambda \leq 7\,\mu m$, $\epsilon_\lambda = 0.35$
$\lambda \geq 7\,\mu m$, $\epsilon_\lambda = 0.15$
The hemispherical spectral values do not change significantly with temperature.

(a) What is the hemispherical, total emissivity of the surface at 1200 K?
(b) If radiation is incident on this metal surface from a blackbody at 6000 K, what is the value of α for the incident radiation?
(c) What is the wavelength $\lambda_{0.5}$ for which 50% of the total radiation emitted by this surface lies in the spectral region $\lambda > \lambda_{0.5}$?
(d) How does the solution to part (c) compare with the wavelength corresponding to maximum radiation for this surface?

3.3 Consider a surface that has the following directional emissivity, ϵ_θ distribution:
$0 \leq \theta \leq 45°$, $\epsilon_\theta = 0.9$
$45° \leq \theta \leq 90°$, $\epsilon_\theta = 0.3$
The surface is isotropic in the ϕ direction. Calculate the ratio of the normal emissivity to the hemispherical emissivity.

3.4 The spectral absorptivity, α_λ and the spectral reflectivity ρ_λ for a spectrally selective, diffuse material vary as follows:
Absorptivity:
$0 \leq \lambda \leq 1.38\ \mu m$, $\alpha_\lambda = 0.2$
$\lambda \geq 1.38\ \mu m$, $\alpha_\lambda = 1.0$
Reflectivity:
$0 \leq \lambda \leq 1.38\ \mu m$, $\rho_\lambda = 0.1$
$\lambda \geq 1.38\ \mu m$, $\rho_\lambda = 0$

(a) Sketch the spectral transmissivity.
(b) If solar radiation, with $G = 750$ W/m^2 and temperature corresponding to a black body at 5800 K, is incident on this material, determine the absorbed, transmitted and reflected fluxes.
(c) If the temperature of this material is 350 K, determine the emissivity, ϵ.
(d) Determine the net radiant heat flux from the material.

3.5 A very large, flat horizontal metal surface (as, for example, a roof) receives solar irradiation of 1150 W/m^2 on its upper surface. The convection heat transfer coefficient on the surface is around 20 W/m$^2 K$. The solar absorptivity of the surface is 0.65, the surface emissivity is 0.15 and the ambient temperature is 30°C. Assume that the bottom of the surface is heavily insulated so that there is no heat transfer from the bottom of the surface. Also, neglect any temperature distribution within the metal surface. For steady state conditions on the metal surface

(a) Determine the temperature of the metal by employing energy balance.
(b) What will be the temperature of the metal if both the emissivity and the absorptivity of the surface are equal to 0.65?
(c) What will be the temperature if both the emissivity and the absorptivity are equal to 0.15? Comment on your results.

3.6 In Problem 3.5, if the surface is assumed to be very highly conducting and there is natural convection from the bottom of the metal surface with a convective heat transfer coefficient of 7 W/m$^2 K$ and the ambient temperature is 35 °C with negligible radiation heat transfer, what will be the surface temperature for conditions corresponding to part (a) of the problem?

3.7 Consider a thin opaque, horizontal plate with an electrical heater on its bottom side. The top side is exposed to ambient air at 25 °C with a convection heat transfer coefficient of 12 W/m^{2}K, solar irradiation of 650 W/m^2 and an effective sky temperature of -40 °C. Determine the electrical power required to maintain the temperature of the surface at 65 °C, if the plate is diffuse and has the following spectral, hemispherical reflectivity

$0 \leq \lambda \leq 2\,\mu m$, $\rho_\lambda = 0.2$

$2\,\mu m \leq \lambda \leq \infty$, $\rho_\lambda = 0.75$

Chapter 4
Radiation Heat Transfer Between Surfaces

A very important topic in radiation is the calculation of radiation heat transfer between surfaces. As engineers, we know that in a typical engineering problem we encounter many surfaces, each having its own temperature, reflectivity, absorptivity, emissivity and so on. These surfaces are usually part of an enclosure or otherwise. The key question to be answered is, What is the net radiative heat transfer from a surface? Even if other modes of heat transfer such as conduction and convection are present, we have to take care of radiation with due diligence. In a combined convection-radiation problem, we may solve the convective heat transfer problem and at every iteration, we may stop and calculate the radiative heat transfer rate, update the convection solver and proceed or we may just have a purely radiation problem.

For example, if we are interested in the cooling of electronics in a satellite, we have a lot of equipment which is generating heat and the temperatures of these have to be controlled. This is done by employing a heat exchanger which will pick up the heat. The fluid which has picked up the heat must be cooled again so that it can be re-circulated to pick up the heat again, as the electronic devices are continuously operating and generating heat. So the hot fluid has to become cold fluid somehow. Therefore, we need a heat exchanger.

Unfortunately, there is no ambient air in outer space, and hence, convective heat transfer is not possible. Therefore, only radiative heat transfer is possible. The design hinges on how we are able to select the surfaces, their configuration, if fins are going to be used, what type of fins are required, their number, thickness, material to be used. We have to solve a combined conduction–radiation problem and design the heat exchanger.

C. Balaji, *Essentials of Radiation Heat Transfer*,
https://doi.org/10.1007/978-3-030-62617-4_4

4.1   Enclosure Theory

There are many applications in which calculation of radiation heat transfer is important, such as satellite temperature control, design of combustion chambers and furnaces, design of radiant superheaters and boilers. Even in other problems as, for
example, the cooling of electronics, radiation also has its part to play as we saw earlier how radiation is significant even at lower temperatures and is comparable with
natural convection. Therefore, it is imperative that we have a method to compute the
radiative heat transfer between surfaces, for which we learn what is called the **enclosure theory**. This was developed by Prof. E. M. Sparrow and his colleagues at the
University of Minnesota in the US in the early 1960s. This enclosure theory, though
developed more than half a century ago, is still in use and has not been challenged.
It is even used by commercial software such as, for example, ANSYS Fluent.

The key idea is like this. Suppose, there is a furnace which has four surfaces with
temperatures and emissivities as shown in Fig. 4.1, radiation from any surface can
fall on any of the other three surfaces.

We account for all the radiation which is originating from a surface and all the
radiation that is falling on the surface and (what is going out = what is coming in)
should be balanced amongst all the surfaces. The system of simultaneous equations
that result upon doing this "accounting" or "book keeping" for every surface can be
solved to obtain the radiative flux (or heat transfer) we desire.

Suppose, we have a configuration as shown (Fig. 4.2), which is called an **open
cavity**, with 3 surfaces and an open top, it is no longer an enclosure. It is like an open
cup. The beauty of the enclosure theory is that we close the top by an imaginary
surface that has zero reflectivity, is a perfect emitter and has a temperature equal to
T_∞. So, we can consider this as the fourth surface and treat the whole geometry as
an enclosure. Thus, any possible configuration on "planet earth" can be treated as an
enclosure!

Fig. 4.1 Radiation exchange
in an enclosure

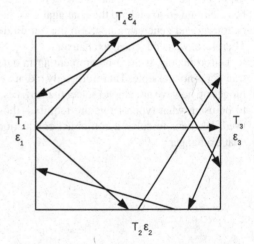

Fig. 4.2 Concept of an imaginary surface under the framework of the enclosure theory

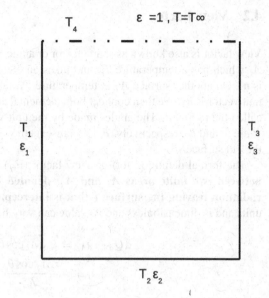

Fig. 4.3 Conversion of a simple one surface problem into one of a two surface enclosure

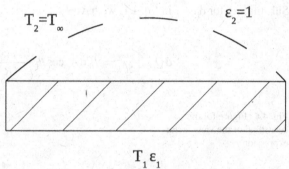

Even if there is a simple one surface enclosure, we can enclose it in a hemispherical basket whose emissivity is 1 and temperature is T_∞ (Fig. 4.3).

Therefore, for any possible configuration, regardless of whether the surfaces are plane, convex or concave, or if some surfaces are open, we can mark a dotted line, and close the geometry. We make everything in the world an enclosure and look at the energy balance of each of the surfaces in the enclosure. This is the key idea behind the enclosure theory.

Even so all of the information we already have through this book, so far, is enough only to calculate the radiation from one surface. But now we are looking at radiation from an enclosure. Obviously, we can see that geometry has a critical role to play. It is intuitively apparent that the size of the various surfaces in the enclosure and the orientation of one surface with respect to the other will eventually decide the net radiation heat transfer from each of the surfaces. Therefore, geometry plays a critical part.

4.2 View Factor

View factor is also known as shape factor or angle factor. Figure 4.4 shows a surface A_i which has a temperature T_i, and takes an elemental area dA_i. The unit vector is n_i. On another surface A_j at temperature T_j, take an elemental area dA_j whose unit vector is n_j. We then connect both elemental area centroids, and this distance is called the radius R. The angles made by the unit vectors n_i and n_j with the radius R are θ_i and θ_j, respectively. F_{i-j} represents the view factor from the ith surface to the jth surface.

The formal definition for the view factor (F_{ij}) is as follows: "**The view factor between two finite areas A_i and A_j, denoted by F_{i-j}, is the fraction of the radiation leaving the surface i that is intercepted by the surface j**". It has no units and is dimensionless and its value can vary between 0 and 1.

$$dQ_{dA_i - dA_j} = I_i.dA_i \cos\theta_i d\omega_{j-i} \tag{4.1}$$

$$d\omega_{j-i} = \frac{dA_j \cos\theta_j}{R^2} \tag{4.2}$$

Substituting for $d\omega_{j-i}$ in Eq. 4.1, we have

$$dQ_{dA_i - dA_j} = I_i.dA_i \cos\theta_i \frac{dA_j \cos\theta_j}{R^2} \tag{4.3}$$

Fig. 4.4 Figure for the derivation of the view factor expression

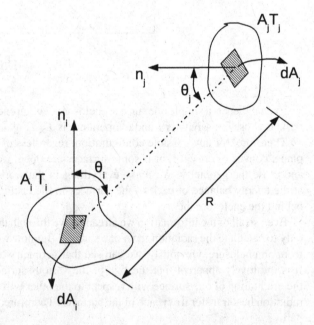

Now, consider i to be a diffuse emitter and a diffuse reflector. This means it does not have a directional preference. The radiation leaving this surface will be the sum of its emission and reflection.

$$J_i = \pi I_{i,e+r} \tag{4.4}$$

where J is called the radiosity, whose units are W/m^2. Substituting for I_i in Eq. 4.3

$$\therefore dQ_{dA_i - dA_j} = \frac{J_i \cos\theta_i \cos\theta_j dA_i dA_j}{\pi R^2} \tag{4.5}$$

The radiation leaving dA_i is $J_i . dA_i$.

The view factor between the two infinitesimal areas (based on our definition) is then given by

$$dF_{dA_i - dA_j} = \frac{dQ_{dA_i - dA_j}}{J_i dA_i} = \frac{\cancel{J_i} \cos\theta_i \cos\theta_j \cancel{dA_i} dA_j}{\cancel{J_i} \cancel{dA_i} \pi R^2} \tag{4.6}$$

$$dF_{dA_i - dA_j} = \frac{\cos\theta_i \cos\theta_j dA_j}{\pi R^2} \tag{4.7}$$

This is a fundamental formula which can be used. For example, if we are computationally very rich, each surface in an enclosure can be divided into thousands of surfaces and this fraction can be calculated for all the thousand surfaces, two at a time. Needless to say, this is computationally expensive and is also unimaginative, to say the least. We will see a little later how we can use algebra to reduce the computational effort associated with evaluating view factors.

Often times, we are not interested in the view factor between elemental areas, but in the view factor between infinitesimal to finite area and then between one finite area and another. The view factor between an infinitesimal area dA_i and finite area A_j is given by

$$F_{dA_i - A_j} = \frac{\int_{A_j} dQ_{dA_i - dA_j}}{J_i dA_i} \tag{4.8}$$

$$F_{dA_i - A_j} = \frac{\cancel{J_i dA_i} \int_{A_j} \cos\theta_i \cos\theta_j dA_j}{\cancel{J_i dA_i} \pi R^2} \tag{4.9}$$

$$F_{dA_i - A_j} = \frac{\int_{A_j} \cos\theta_i \cos\theta_j dA_j}{\pi R^2} \tag{4.10}$$

Finally, the view factor between two finite areas A_i and A_j denoted by $F_{A_i-A_j}$ or simply F_{i-j} is given by

$$F_{A_i-A_j} = F_{i-j} = \frac{\int_{A_j}\int_{A_i} dQ_{dA_i-dA_j}}{\int J_i dA_i} \tag{4.11}$$

Substituting for the numerator from Eq. 4.5 together with the assumption of uniform radiosity wherein J_i can be pulled out of both the numerator and denominator, we have

$$F_{A_i-A_j} = \frac{1}{A_i}\left(\frac{\int_{A_j}\int_{A_i}\cos\theta_i\cos\theta_j dA_i dA_j}{\pi R^2}\right) \tag{4.12}$$

By induction,

$$F_{A_j-A_i} = \frac{1}{A_j}\left(\frac{\int_{A_i}\int_{A_j}\cos\theta_i\cos\theta_j dA_i dA_j}{\pi R^2}\right) \tag{4.13}$$

Now, if we want to solve a four zone enclosure problem, first we need view factors. If we were to integrate and get all the view factors, evidently, it is going to take a lot of time. So, we need to see if there are some clever ways of getting the view factor. This whole subfield, where we try to manipulate algebraically to get the view factors, with minimum recourse to the original formula involving integrals, is called **"view factor algebra"**. If we are computationally very rich, we can write programs using the above formula to get the view factors, as already mentioned. But for simple surfaces, can we do better and use a simpler approach to get the view factor? From Eqs. (4.12) and (4.13), it evident that

$$A_i F_{ij} = A_j F_{ji} \tag{4.14}$$

Equation 4.14 is known as the reciprocal rule or reciprocal relation and is our first ammunition in the view factor algebra arsenal!

4.3 View Factor Algebra

Consider an enclosure of N sides, as shown in Fig. 4.5. There will be N^2 view factors, associated with this enclosure with N being 5 in this specific case. The view factors can be written out as a matrix, as given below.

Fig. 4.5 A typical N surface
enclosure (N = 5)

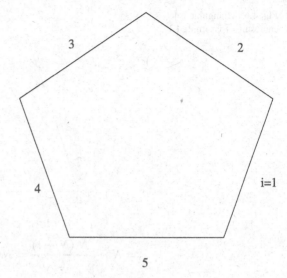

$$\text{View factor matrix} = \begin{bmatrix} F_{11} & F_{12} & \cdots\cdots\cdots & F_{1N} \\ F_{21} & F_{22} & \cdots\cdots\cdots & F_{2N} \\ \vdots & \vdots & & \vdots \\ F_{N1} & F_{N2} & \cdots\cdots\cdots & F_{NN} \end{bmatrix}$$

Is it a good idea to get the 25 view factors for this 5 sided enclosure using the integration method? For the ith surface, common sense tells us the following,

$$\sum_{j=1}^{N} F_{ij} = 1 \qquad \text{for all } i \tag{4.15}$$

This is called the **summation rule**. The sum has to be 1 because this follows energy balance. For the 5 surface enclosure under consideration, for the surface 1, we can write Eq. (4.15) as

$$F_{11} + F_{12} + F_{13} + F_{14} + F_{15} = 1 \tag{4.16}$$

For an N surface enclosure, N such rules are available. We already saw that $A_i F_{ij} = A_j F_{ji}$. For an N surface enclosure there are NC_2 such **reciprocal rules**.

$$\text{Number of reciprocal rules} = NC_2 = \frac{N.(N-1)}{2} \tag{4.17}$$

We can exploit the summation and reciprocal rules and, therefore, the number of independent view factors to be determined for a N surface enclosure is

Fig. 4.6 Triangular
enclosure of Example 4.1

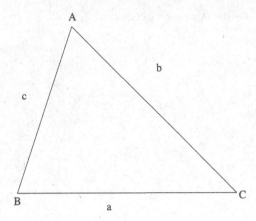

$$= N^2 - \frac{N.(N-1)}{2} = NC_2 \qquad (4.18)$$

If all the surfaces are plane or convex, the self view factors $F_{ii} = 0$. Therefore, if all the surfaces are plane or convex, then the total number of view factors to be independently evaluated $= NC_2 - N = \dfrac{N(N-1)}{2} - N$. A key goal of view factor algebra is to determine the number of independent view factors which have to be evaluated necessarily by adopting the fundamental view factor integral, i.e. Eq. 4.12.

Example 4.1 Consider a two-dimensional evacuated triangular enclosure with three surfaces of length a, b and c (Fig. 4.6). Determine all the view factors.

Solution: $\therefore$ Total number of view factors $= N^2 = 9$
Sum rules = 3; reciprocal rules $= 3C_2 = 3$; Self view factors = 3
View factors to be obtained independently $= 9 - 9 = 0$
So, there is no need to work out any view factor using the integration formula. Having obtained guidance from view factor algebra, we now set out to determine all the view factors purely by algebra.

$$\text{Area of BC (a)} = a \times 1 = a \left(\text{m}^2/\text{m}\right)$$
$$\text{Area of AC (b)} = b \times 1 = b \left(\text{m}^2/\text{m}\right)$$
$$\text{Area of AB (c)} = c \times 1 = c \left(\text{m}^2/\text{m}\right)$$

Using three sum rules

$$F_{aa} + F_{ab} + F_{ac} = 1 \qquad (4.19)$$
$$F_{ba} + F_{bb} + F_{bc} = 1 \qquad (4.20)$$
$$F_{ca} + F_{cb} + F_{cc} = 1 \qquad (4.21)$$

All self view factors are zero, $F_{aa} = F_{bb} = F_{cc} = 0$.

$$\cancel{F_{aa}} + F_{ab} + F_{ac} = 1 \tag{4.22}$$

$$F_{ba} + \cancel{F_{bb}} + F_{bc} = 1 \tag{4.23}$$

$$F_{ca} + F_{cb} + \cancel{F_{cc}} = 1 \tag{4.24}$$

$$F_{ab} + F_{ac} = 1 \tag{4.25}$$
$$F_{ba} + F_{bc} = 1 \tag{4.26}$$
$$F_{ca} + F_{cb} = 1 \tag{4.27}$$

Multiplying the first equation by a, the second by b and the third by c,

$$a F_{ab} + a F_{ac} = a \tag{4.28}$$
$$b F_{ba} + b F_{bc} = b \tag{4.29}$$
$$c F_{ca} + c F_{cb} = c \tag{4.30}$$

Now, we add Eqs. 4.28 and 4.29, subtract (4.30) from the sum, followed by an application of the reciprocal rule and upon doing this we get the following

$$2a F_{ab} = a + b - c \tag{4.31}$$

[To make things more explicit, while doing the above manipulation, we have used the following relations

$$a F_{ab} = b F_{ba}, \ a F_{ac} = c F_{ca} \text{ and } b F_{bc} = c F_{cb}]$$

$$\therefore \ F_{ab} = \frac{a + b - c}{2a} \tag{4.32}$$

For checking if this result is correct, consider an equilateral triangle. All view factors are 0.5 which is intuitively apparent.

Example 4.2 Consider a two-dimensional V-groove or wedge with surfaces 1 and 2 as shown in Fig. 4.7, and whose interior angle is α. Determine F_{12}.

Solution:
Since we are using enclosure theory, the first step is to mark a dotted line and close this, making the wedge or groove an enclosure.
We determine the length of side 3 using the sine function (Fig. 4.8).
Length of side 3 $= 2 \times L \sin(\alpha/2)$

$$F_{12} = \frac{L_1 + L_2 - L_3}{2L_1} \tag{4.33}$$

$$F_{12} = \frac{L + L - 2L \sin(\alpha/2)}{2L} \tag{4.34}$$

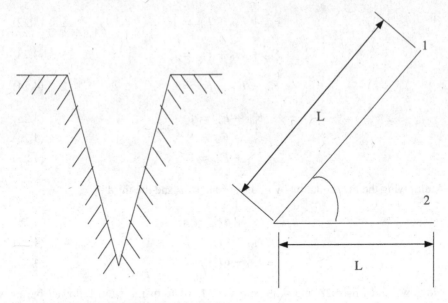

Fig. 4.7 Wedge type enclosure (Example 4.2)

Fig. 4.8 View factors in a
wedge (Example 4.2)

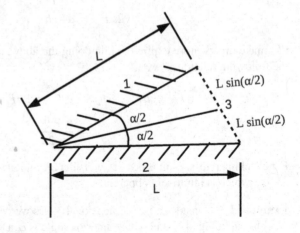

$$F_{12} = \frac{2L - 2L\sin(\alpha/2)}{2L} \tag{4.35}$$

$$F_{12} = \frac{2\!\!\!/L(1 - \sin(\alpha/2))}{2\!\!\!/L} \tag{4.36}$$

As an engineer, how will we validate this result?

If $\alpha = 180°$, the wedge will open completely. Then the two sides do not see each other and there is no interaction between them. Therefore,

Fig. 4.9 View factors in a concentric pipe enclosure (Example 4.3)

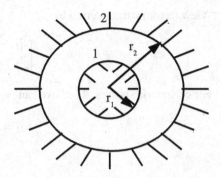

$$sin(\alpha/2) = sin(90) = 1 \tag{4.37}$$
$$F_{12} = 1 - sin(\alpha/2) = 0 \tag{4.38}$$

Equation (4.38) is consistent with our common sense understanding of the situation.

Example 4.3 Consider two concentric pipes where the inner pipe is carrying a fluid (Fig. 4.9). The radius of the inner pipe is r_1 while that of the outer pipe is r_2. The pipes are infinitely deep, perpendicular to the cross-section. Do not consider the radiation from the outer side of the external pipe and the inner side of the inner pipe. For this cylindrical duct, get all the view factors. Assume that there is vacuum between the two pipes.

Solution:
The view factor matrix will be

$$\begin{bmatrix} F_{11} & F_{12} \\ F_{21} & F_{22} \end{bmatrix} \tag{4.39}$$

$$F_{11} = 0 \tag{4.40}$$

(because surface 1 is a convex surface and so the self view factor is 0)

$$F_{12} = 1 \tag{4.41}$$
$$A_1 F_{12} = A_2 F_{21} \tag{4.42}$$
$$F_{21} = A_1/A_2 \tag{4.43}$$
$$F_{22} = 1 - A_1/A_2 \tag{4.44}$$

Concentric cylinders:

$$F_{21} = r_1/r_2 \tag{4.45}$$
$$F_{22} = 1 - r_1/r_2 \tag{4.46}$$

View factor matrix is given by

$$\begin{bmatrix} 0 & 1 \\ \frac{r_1}{r_2} & 1 - \frac{r_1}{r_2} \end{bmatrix} \tag{4.47}$$

For concentric spheres, the above can be extended as follows:

$$F_{21} = r_1^2 / r_2^2 \tag{4.48}$$

$$F_{22} = 1 - \frac{r_1^2}{r_2^2} \tag{4.49}$$

The view factor matrix is given by

$$\begin{bmatrix} 0 & 1 \\ \frac{r_1^2}{r_2^2} & 1 - \frac{r_1^2}{r_2^2} \end{bmatrix} \tag{4.50}$$

Getting back to the example, in practical applications, this is one way of insulation and is used in transporting liquid nitrogen and oxygen over large distances. The temperature of the liquid will be 150 K and the outside may be at 300 K. There will be a heat leak causing the liquid nitrogen to vaporize. So, we have a small vent to allow it to escape. We can work out the rate of vaporization of this and if we know the transportation time, we can calculate the amount of liquid nitrogen that will be left at the receiving end.

Example 4.4 Consider a two-dimensional quadrilateral enclosure as shown in Fig. 4.10. Determine F_{ac}.

Fig. 4.10 Quadrilateral enclosure (Example 4.4)

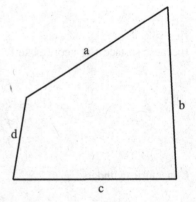

Fig. 4.11 Depiction of Hottel's crossed string method for a quadrilateral enclosure

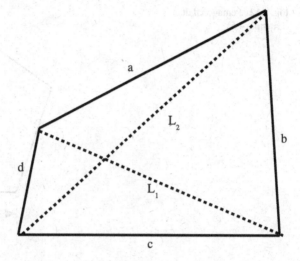

Solution:

Total number of view factors $= 4^2 = 16$

Sum rules $= 4$

Reciprocal rules $= 4C_2 = 6$

Self view factors $= 4$

Number of view factors to be independently determined $= 16 - 14 = 2$ (F_{ac} and F_{bd})

If F_{ac} is obtained, we can use the same formula for F_{bd}.

The first step is to complete the two diagonals L_1 and L_2 (Fig. 4.11)

$$F_{aa} + F_{ab} + F_{ac} + F_{ad} = 1 \qquad (4.51)$$

$$\therefore F_{ac} = 1 - (F_{ab} + F_{ad}) \qquad (4.52)$$

$$F_{ab} = \frac{a + b - L_1}{2a} \qquad (4.53)$$

$$F_{ad} = \frac{a + d - L_2}{2a} \qquad (4.54)$$

Substituting for F_{ab} and F_{ad} in Eq. 4.52

$$F_{ac} = 1 - \frac{(a + d - L_1) + (a + d - L_2)}{2a} \qquad (4.55)$$

$$F_{ac} = 1 - \frac{2a + b + d - (L_1 + L_2)}{2a} \qquad (4.56)$$

Fig. 4.12 Pentagonal duct
(Example 4.5)

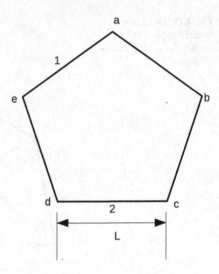

$$F_{ac} = \frac{(L_1 + L_2) - (b + d)}{2a} \qquad (4.57)$$

This is called the **Hottel's Crossed String** method. This is because if we look at the formula, we find that the view factor between surfaces a and c is the sum of the crossed strings $(L_1 + L_2)$ minus the sum of the uncrossed strings $(b + d)$ divided by two times the line segment a.

$$\therefore F_{ac} = (\text{sum of the crossed strings} - \text{sum of the uncrossed strings})/2a$$

This is a very powerful expression for two-dimensional enclosures, which can be used for any geometry.

Example 4.5 Consider a two-dimensional regular pentagonal duct as shown in Fig. 4.12. Determine F_{12}.

Solution:
Exterior angle = $360°/n = 360°/5 = 72°$
So the interior angle is $108°$. Let us draw the diagonals as shown in Fig. 4.13.
Using Hottel's Crossed string method

$$F_{12} = \frac{[ad + ce - (ac + de)]}{2ae} \qquad (4.58)$$

$$ad = \frac{L/2}{\cos 72} = 1.618L \qquad (4.59)$$

$$F_{12} = \frac{2 \times 1.618L - (L + 1.618L)}{2L} = 0.309 \qquad (4.60)$$

Fig. 4.13 Pentagonal duct
with depiction of
crossed-string method
(Example 4.5)

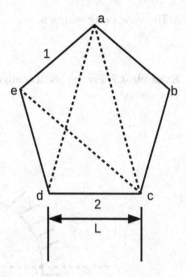

Example 4.6 Consider an infinitely deep semi-circular disc of radius r and unit depth together with the base. Get all the view factors for this geometry (Fig. 4.14).

Fig. 4.14 Semi-circular disk
(Example 4.6)

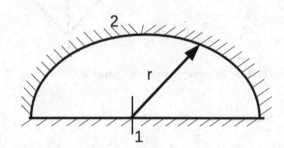

Solution:

$$F_{11} + F_{12} = 1 \qquad (4.61)$$

$F_{11} = 0; F_{12} = 1$

$$A_1 F_{12} = A_2 F_{21} \qquad (4.62)$$
$$2r \times 1 = \pi r \times F_{21} \qquad (4.63)$$
$$F_{21} = \frac{2}{\pi} \qquad (4.64)$$
$$F_{22} = 1 - F_{21} \qquad (4.65)$$
$$F_{22} = 1 - \frac{2}{\pi} \qquad (4.66)$$

∴ The view factor matrix is

$$\begin{bmatrix} 0 & 1 \\ \frac{2}{\pi} & 1 - \frac{2}{\pi} \end{bmatrix}$$

Example 4.7 For the given configuration (see Fig. 4.15) with unit depth, get all the view factors

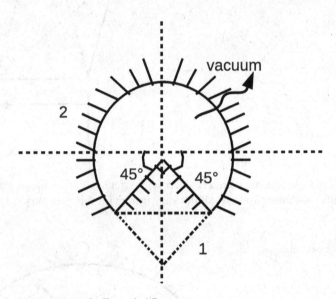

Fig. 4.15 Problem geometry for Example 4.7

Solution

$$F_{11} + F_{12} = 1 \tag{4.67}$$

$F_{11} = 0; \ F_{12} = 1$

$$A_1 F_{12} = A_2 F_{21} \tag{4.68}$$

$$2r \times 1 = 2\pi r \times \frac{3}{4} \times F_{21} = \frac{3\pi r}{2} F_{21} \tag{4.69}$$

$$F_{21} = \frac{4}{3\pi} \tag{4.70}$$

$$F_{22} = 1 - \frac{1}{\frac{3}{4}\pi} \tag{4.71}$$

∴ The view factor matrix is

$$\begin{bmatrix} 0 & 1 \\ \frac{4}{3\pi} & 1 - \frac{4}{3\pi} \end{bmatrix}$$

4.4 View Factors From Direct Integration

Example 4.8 Determine the view factor F_{d1-2} between a differential area (dA_1) and a finite disk of radius r_o at a height H (Fig. 4.16).

Solution:
The first thing to recognize in this problem is that the tricks of view factor algebra which we learned thus far will not be of much help to us in solving this problem. We have to take recourse to the fundamental view factor expression.

$$F_{dA_1-A_2} \text{ or } F_{d1-2} = \int_{A_2} \frac{\cos \theta_i \cos \theta_j dA_2}{\pi R^2} \qquad (4.72)$$

We have to convert the problem into one in which the variable is r, which can take a value between 0 and r_o. Therefore, first, we have to take an infinitesimally small area on the disc, take a thickness of dr so that the area will be $2\pi r dr$. The area dA_2, can be replaced by $2\pi r dr$. Then if it is at a radius r, from fundamental trigonometric principles, we can find out what R is, (Please note that R $\neq$ H). Then the two terms $\cos \theta_i$ and $\cos \theta_j$ also have to be represented in terms of known quantities. After all this, the integration can be done easily.

$$\theta_i = \theta_j = \theta \qquad (4.73)$$

Fig. 4.16 View factor between a differential area and a disk of finite area

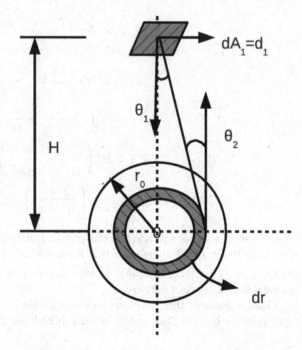

$$R^2 = r^2 + H^2 \tag{4.74}$$

$$\cos\theta = H/R \tag{4.75}$$

$$dA_2 = 2\pi r\, dr \tag{4.76}$$

Substituting for $\cos\theta$, R and dA_2 in the integral,

$$F_{d1-2} = \int_0^{r_0} \frac{H^2}{R^2} \frac{2\pi r\, dr}{\pi R^2} \tag{4.77}$$

$$F_{d1-2} = 2H^2 \int_0^{r_0} \frac{r\, dr}{R^4} \tag{4.78}$$

$$F_{d1-2} = 2H^2 \int_0^{r_0} \frac{r\, dr}{(r^2 + H^2)^2} \tag{4.79}$$

$$Let,\ r^2 + H^2 = y,\ 2r\, dr = dy \tag{4.80}$$

$$Limits,\ r = 0,\ y = H^2;\ r = r_0,\ y = H^2 + r_0^2,\ dr = \frac{dy}{2r} \tag{4.81}$$

$$F_{d1-2} = 2H^2 \int_{H^2}^{H^2+r_0^2} \frac{\frac{dy}{2r}\, r}{y^2} \tag{4.82}$$

$$F_{d1-2} = H^2 \int_{H^2}^{H^2+r_0^2} \frac{dy}{y^2} \tag{4.83}$$

$$F_{d1-2} = H^2 \left(\frac{-1}{y}\right)_{H^2}^{H^2+r_0^2} \tag{4.84}$$

$$F_{d1-2} = H^2 \left(\frac{-1}{H^2 + r_0^2} + \frac{1}{H^2}\right) \tag{4.85}$$

$$F_{d1-2} = H^2 \left(\frac{H^2 - H^2 + r_0^2}{(H^2 + r_0^2)H^2}\right) \tag{4.86}$$

$$F_{d1-2} = \left(\frac{r_0^2}{H^2 + r_0^2}\right) \tag{4.87}$$

A good question to ask now is What is the physical interpretation of the solution to this problem? When H is infinite, $F = 0$. If $r_0/H \ll 1$, we can reduce the expression to $F_{d1-2} = \frac{r_0^2}{H^2}$ and it confirms that the view factor varies as the inverse of the square of the distance between the two.

Please remember that since one area was infinitesimal, the integration was not only possible but was also straightforward. If both are finite areas and one is not a disc, the resulting mathematics would be very tedious. Many groups of people in the world in the 1960–1980s worked on developing view factor relations and solving these integrals, which was considered a very important activity. The view factors are repeatedly used in radiosity calculations in problems involving radiation alone and more so in those concerning multi-mode heat transfer. View factors are also one of the reasons why double precision was required. We need view factors up to the seventh or eighth decimal accuracy, as, if we have a hundred thousand view factors, we cannot round off each to the second or third decimal. When we are doing a convective solver or convection and radiation together, such an approximation will lead to a lot of errors.

But now, with computational resources becoming more powerful and more programs being available, this activity of research is not so prominent these days. These are mostly considered to be well settled problems.

So far, we have restricted our working to two-dimensional surfaces. However, most of the surfaces in reality are three-dimensional surfaces, and the view factors cannot be obtained by just algebra. Researchers have developed elaborate techniques for this, of which one of the most important is **contour integration**. If we look at dA_1 and dA_2 we see that they can be written as $dA_1 = dx_1dy_1$ and $dA_2 = dx_2dy_2$.

Therefore, 4 integrations are involved. Elemental strips on A_1 and A_2 are considered, $\cos\theta_i$ and $\cos\theta_j$ are determined and four integrations are performed. These are available in the form of charts and Tables. These are given in Figs. 4.17, 4.18, 4.19 and 4.20 for three commonly encountered geometries namely two parallel rectangles, two perpendicular rectangles with a common edge and two parallel coaxial circular disks. We will now solve a few problems using these charts.

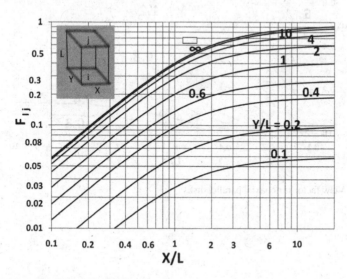

Fig. 4.17 View factor for aligned parallel rectangles

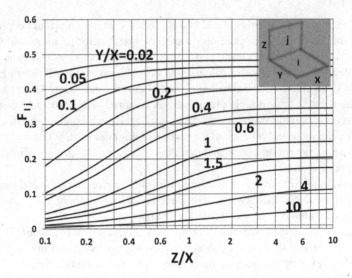

Fig. 4.18 View factor for perpendicular rectangles with a common edge

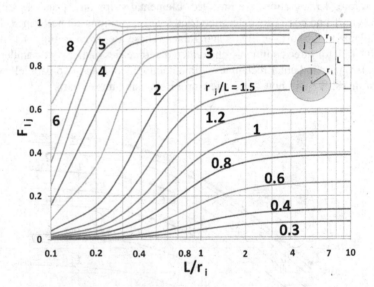

Fig. 4.19 View factor for coaxial parallel disks

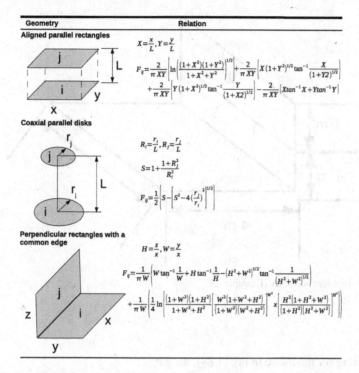

Fig. 4.20 View factors for a few frequently encountered three-dimensional geometries

Example 4.9 Consider two perpendicular rectangles as shown in Fig. 4.21. Determine the view factor F_{1-2}.

Solution:
Let us learn about the decomposition rule first. It says that

$$F_{2-13} = F_{2-1} + F_{2-3} \tag{4.88}$$

Equation 4.88 is a consequence of simple energy balance. Whatever radiation is originating from 2 and is falling on the combined area (13) must be equal to the sum of that originating from 2 and falling on 1 and 3 individually. We know that

$$A_2 F_{2-13} = A_{13} F_{13-2} \tag{4.89}$$

Further

$$A_1 F_{12} = A_2 F_{21} \tag{4.90}$$

and

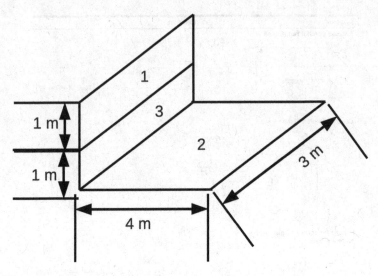

Fig. 4.21 Problem geometry (Example 4.9)

$$A_2 F_{2-3} = A_3 F_{3-2} \qquad (4.91)$$

Substituting for the above in Eq. (4.88), we get

$$\frac{A_{13} F_{13-2}}{A_2} = \frac{A_1}{A_2} F_{12} + \frac{A_3}{A_2} F_{32} \qquad (4.92)$$

From Eq. 4.92, it is clear that

$$F_{13-2} \neq F_{1-2} + F_{3-2} \qquad (4.93)$$

Getting back to the problem under consideration
  $z/x = 0.67$;
  $y/x = 4/3 = 1.333$
  $F_{2-13} = 0.14$ (from the chart)

$$F_{2-13} = F_{2-1} + F_{2-3} \qquad (4.94)$$

$F_{2-3} = 0.07$ (from the chart $z/x = 0.33$; $y/x = 4/3 = 1.333$)

$$0.14 = F_{2-1} + 0.07 \qquad (4.95)$$
$$F_{2-1} = 0.07 \qquad (4.96)$$
$$A_1 F_{12} = A_2 F_{21} \qquad (4.97)$$

$A_1 = 3 \text{ m}^2$; $A_2 = 12 \text{ m}^2$

Fig. 4.22 Problem geometry
(Example 4.10)

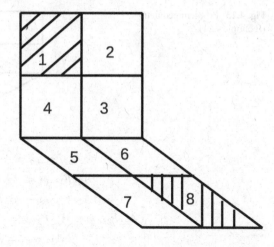

$$\therefore \; 3 \times F_{12} = 12 \times 0.07 \tag{4.98}$$
$$\text{and so} \;\; F_{12} = 0.28 \tag{4.99}$$

Example 4.10 In the figure given (Fig. 4.22) get F_{1-8}

Solution:
We have to use the law called "law of corresponding corners", according to which
$F_{18} = F_{27}$

$$A_{1234}F_{1234-5678} = A_1 F_{1-5678} + A_2 F_{2-5678} + A_3 F_{3-5678} + A_4 F_{4-5678} \tag{4.100}$$
$$A_{1234}F_{1234-5678} = A_{12}F_{12-5678} + A_{34}F_{34-5678} \tag{4.101}$$
$$A_{1234}F_{1234-5678} - A_{34}F_{34-5678} = A_{12}F_{12-5678} \tag{4.102}$$
$$A_{12}F_{12-5678} = A_1 F_{1-5678} + A_2 F_{2-5678} \tag{4.103}$$
$$F_{1-5678} = F_{1-567} + F_{1-8} \tag{4.104}$$
$$F_{2-5678} = F_{2-568} + F_{1-8} \tag{4.105}$$
$$\text{But} \quad F_{1-8} = F_{2-7} \tag{4.106}$$
$$F_{12-5678} = 2.A_1 F_{1-8} + A_{12}F_{12-56} + A_1 F_{1-7} + A_2 F_{2-8} \tag{4.107}$$
$$A_{1234}F_{1234-56} = A_{12}F_{12-56} + A_{34}F_{34-56} \tag{4.108}$$
$$A_{1234}F_{1234-56} - A_{34}F_{34-56} = A_{12}F_{12-56} \tag{4.109}$$
$$A_{14}F_{14-57} = A_{14}F_{14-5} + A_{14}F_{14-7} \tag{4.110}$$
$$A_{14}F_{14-57} - A_{14}F_{14-5} = A_1 F_{1-7} + A_4 F_{4-7} \tag{4.111}$$
$$A_{14}F_{14-57} - A_{14}F_{14-5} - A_4 F_{4-7} = A_1 F_{1-7} \tag{4.112}$$
$$A_4 F_{4-7} + A_4 F_{4-5} = A_4 F_{4-57} \tag{4.113}$$
$$A_4 F_{4-7} = A_4 F_{4-57} - A_4 F_{4-5} \tag{4.114}$$

Fig. 4.23 Problem geometry
(Example 4.11)

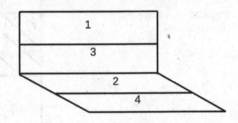

$$A_{23}F_{23-68} = A_2 F_{2-68} + A_3 F_{3-68} \qquad (4.115)$$

$$A_{23}F_{23-68} = A_{23} F_{23-6} + A_{23} F_{23-8} \qquad (4.116)$$

$$A_{23}F_{23-68} - A_{23} F_{23-6} = A_{23} F_{23-8} \qquad (4.117)$$

$$A_{23}F_{23-68} - A_{23} F_{23-6} = A_2 F_{2-8} + A_3 F_{3-8} \qquad (4.118)$$

$$A_3 F_{3-68} = A_3 F_{3-6} + A_3 F_{3-8} \qquad (4.119)$$

$$A_3 F_{3-68} - A_3 F_{3-6} = A_3 F_{3-8} \qquad (4.120)$$

$$A_2 F_{2-8} = A_{23} F_{23-68} - A_{23} F_{23-6} - A_3 F_{3-68} + A_3 F_{3-6} \qquad (4.121)$$

$$A_1 F_{1-7} = A_{14} F_{14-57} - A_{14} F_{14-5} - A_4 F_{4-57} + A_4 F_{4-5} \qquad (4.122)$$

$$F_{1-8} = \frac{1}{2.A_1}(A_{12} F_{12-5678} - A_{12} F_{12-56} - A_1 F_{1-7} - A_2 F_{2-8}) \qquad (4.123)$$

$$F_{1-8} = \frac{1}{2.A_1}(A_{1234} F_{1234-5678} - A_{34} F_{34-5678} - A_{1234} F_{1234-56}$$

$$+ A_{34} F_{34-56} - A_{14} F_{14-57} + A_{14} F_{14-5} + A_4 F_{4-57} - A_4 F_{4-5}$$

$$- A_{23} F_{23-68} + A_{23} F_{23-6} + A_3 F_{3-68} - A_3 F_{3-6}) \qquad (4.124)$$

Example 4.11 For the geometry given in Fig. 4.23, determine F_{14}

Solution:
We know that

$$A_{13} F_{13-24} = A_1 F_{1-24} + A_3 F_{3-24} \qquad (4.125)$$

$$\therefore \; A_{13} F_{13-24} - A_3 F_{3-24} = A_1 F_{1-24} \qquad (4.126)$$

Further, $\quad A_1 F_{1-24} = A_1 F_{1-2} + A_1 F_{1-4} \quad$ (decomposition rule) $\qquad (4.127)$

Substituting Eq. (4.126) in Eq. (4.125),

$$A_{13} F_{13-24} - A_3 F_{3-24} = A_1 F_{1-2} + A_1 F_{1-4} \qquad (4.128)$$

$$F_{1-4} = \frac{1}{A_1}(A_{13} F_{13-24} - A_3 F_{3-24} - A_1 F_{1-2}) \qquad (4.129)$$

$$A_{13} F_{13-2} = A_1 F_{1-2} + A_3 F_{3-2} \qquad (4.130)$$

Fig. 4.24 Problem geometry
(Example 4.12)

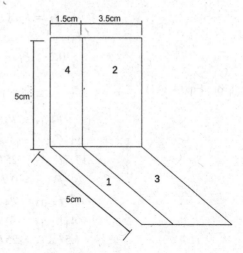

$$A_{13}F_{13-2} - A_3F_{3-2} = A_1F_{1-2} \tag{4.131}$$

$$A_{13} = A_1 + A_3 \tag{4.132}$$

Substituting Eq. (4.131) in Eq. (4.129),

$$F_{1-4} = \frac{1}{A_1}(A_{13}F_{13-24} - A_3F_{3-24} - A_{13}F_{13-2} + A_3F_{3-2}) \tag{4.133}$$

Example 4.12 Consider the configuration shown in Fig. 4.24. Two parallel rectangles with a common edge are further subdivided into 2 rectangles each so that we have 4 areas A_1, A_2, A_3 and A_4. Determine the view factor F_{12}.

Solution:

$$A_1 = 5 \times 1.5 = 7.5 \text{ m}^2 \tag{4.134}$$

$$A_4 = 7.5 \text{ m}^2 \tag{4.135}$$

$$A_3 = 5 \times 3.5 = 17.5 \text{ m}^2 \tag{4.136}$$

$$A_2 = 17.5 \text{ m}^2 \tag{4.137}$$

$$A_{13} = 25 \text{ m}^2; A_{24} = 25 \text{ m}^2 \tag{4.138}$$

F_{13-24}(can be obtained from the chart) $= 0.2 = F_{24-13}$;
$y/x = 1; z/x = 1$
F_{14}(from the chart) $= 0.12 = F_{41}$ ($y/x = 3.33; z/x = 3.33$)
F_{23}(from the chart) $= 0.17 = F_{32}$
($y/x = 5/3.5 = 1.428; z/x = 5/3.5 = 1.428$)
Now, we have to use the reciprocal rule and decomposition rule and manipulate them algebraically to get the remaining view factors.

$$F_{2-13} = F_{2-1} + F_{2-3} \tag{4.139}$$
$$F_{13-24} = F_{13-2} + F_{13-4} \tag{4.140}$$
$$0.2 = F_{13-2} + F_{13-4} \tag{4.141}$$

From Eq. (4.141)

$$F_{2-13} = F_{2-1} + 0.17 \tag{4.142}$$
$$A_2 F_{2-13} = A_{13} F_{13-2} \tag{4.143}$$
$$17.5 F_{2-13} = 25 F_{13-2} \tag{4.144}$$
$$F_{13-2} = 0.7 F_{2-13} \tag{4.145}$$
$$F_{4-13} = F_{4-1} + F_{4-3} \tag{4.146}$$
$$A_4 F_{4-13} = A_{13} F_{13-4} \tag{4.147}$$
$$7.5 F_{4-13} = 25 F_{13-4} \tag{4.148}$$
$$F_{13-4} = 0.3 F_{4-13} \tag{4.149}$$

Substituting for F_{13-4} and F_{13-2} into Eq. (4.141),

$$0.2 = 0.7(F_{2-1} + 0.17) + 0.3(0.12 + F_{4-3}) \tag{4.150}$$

But, $F_{3-4} = F_{2-1}$ by the law of corresponding corners.
   On simplifying, we get

$$0.0415 = 0.7 F_{2-1} + 0.3 F_{12} \tag{4.151}$$
$$A_1 F_{12} = A_2 F_{21} \tag{4.152}$$
$$17.5 F_{12} = 7.5 F_{2-1} \tag{4.153}$$

$$F_{21} = 0.428 F_{12} \tag{4.154}$$
$$0.0415 = 0.7 \times 0.428 F_{12} + 0.3 F_{12} \tag{4.155}$$
$$0.0415 = 0.5996 F_{12} \tag{4.156}$$
$$F_{12} = 0.07 \tag{4.157}$$

Because of errors associated with reading the charts, the final result may vary from 0.06 to 0.08. Furthermore, the error will be more when the view factor is much smaller.

Example 4.13 An enclosure is in the shape of the frustum of a cone. The dimensions of the enclosure are given in Fig. 4.25. Determine all the view factors.

Solution:
If an enclosure problem is given, it is always a good idea to state the number of view factors to be determined, the number of sum rules, reciprocal rules, self view factors and then find out how many view factors have to be independently determined. By

Fig. 4.25 Problem geometry
(Example 4.13)

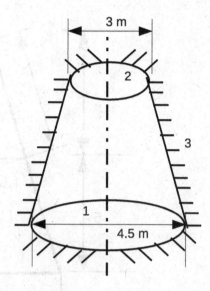

this, we will get an idea of whether we have to use the charts or by manipulation alone, the problem of view factor determination can be solved.

Total number of view factors $= N^2 = 9$

Sum rules $= 3$

Reciprocal rules $= 3$

Self view factors $= 2$

Total $= 8$

So only one view factor (F_{12}) has to be independently determined, and this can be got from the chart.

$r_i = 2.25$ m; $r_j = 1.5$ m; $L = 5$ m; $L/r_i = 2.22$ and $r_j/L = 0.33$

From the chart (Fig. 4.19), $F_{12} = 0.07$

From the sum rule for surface 1

$$F_{11} + F_{12} + F_{13} = 1 \tag{4.158}$$

$$\text{Also } F_{11} = 0 \tag{4.159}$$

$$F_{13} = 1 - 0.07 = 0.93 \tag{4.160}$$

$$A_1 F_{12} = A_2 F_{21} \tag{4.161}$$

$$F_{21} = (A_1/A_2) F_{12} \tag{4.162}$$

$$= (r_1/r_2)^2 F_{12}$$

$$= 0.157 \tag{4.163}$$

From the sum rule for surface 2 $\tag{4.164}$

$$F_{21} + F_{22} + F_{23} = 1 \tag{4.165}$$

$$F_{22} = 0;$$

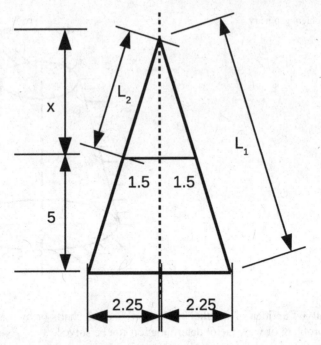

Fig. 4.26 Use of similar triangles principle in Example 4.13

$$\therefore \quad F_{23} = 1 - F_{21} = 0.843 \tag{4.166}$$
$$A_1 F_{13} = A_3 F_{31} \tag{4.167}$$
$$A_2 F_{23} = A_3 F_{32} \tag{4.168}$$

To determine A_3, we have to take recourse to basic geometry.
From law of similarity of triangles (see Fig. 4.26),

$$\frac{x}{1.5} = \frac{x+5}{2.25} \tag{4.169}$$
$$x = 10 \tag{4.170}$$

Surface area of the cone $= \pi r L$, where L-slant length

$$L_1 = \sqrt{225 + 5.0625} = 15.16 \text{ m} \tag{4.171}$$
$$L_2 = \sqrt{100 + 2.25} = 10.1 \text{ m} \tag{4.172}$$

Surface area of the frustum $A_3 = \pi(r_1 L_1 - r_2 L_2) = 59.6 \text{ m}^2$
Substituting this in Eq. (4.167), we get

$$A_1 F_{13} = A_3 F_{31} \tag{4.173}$$

$$15.9 \times 0.93 = 59.6 \times F_{31} \tag{4.174}$$

$$F_{31} = 0.248 = 0.25 \tag{4.175}$$

From Eq. (4.168), $A_2 F_{23} = A_3 F_{32}$

$$7.06 \times 0.843 = 59.6 \times F_{32} \tag{4.176}$$

$$F_{32} = 0.10 \tag{4.177}$$

$$F_{33} = 1 - (F_{31} + F_{32}) = 0.65 \tag{4.178}$$

$\therefore$ The final view factor matrix for the above configuration is

$$\begin{bmatrix} 0 & 0.07 & 0.93 \\ 0.157 & 0 & 0.843 \\ 0.25 & 0.10 & 0.65 \end{bmatrix}$$

The above problem involved only a 3×3 matrix, and so, we are able to solve it by hand. If we encounter a 10×10 or a 20×20 matrix (meaning a 10 or 20 surface enclosure), we will need to write a computer program to determine the view factor matrix and store it once and for all. When the temperatures are dynamically updated, this matrix will not change, as it depends only on the geometry. However, if we have a problem, where the surfaces are also changing and the dimensions are also changing, which can happen if we have an ablating surface like in the case of a re-entry vehicle entering the earth's atmosphere, where a few millimeters of the surface may sublimate because of heat, the geometry itself will change and geometry linked view factor updation should be done in such a case.

So far, we have developed the background to do the radiation analysis. The next logical question to ask is: If there is an n-surface enclosure, how do we find out the net radiation heat transfer between the various surfaces? We will answer this question in the ensuing sections.

4.5 Enclosure Analysis

There are several ways of solving problems involving radiation from multiple sur-faces. In undergraduate courses, we would have done something called network analysis, in which the radiation resistance will be drawn and using series and parallel combinations, the analogy to electrical resistance will be used to solve the problem. The major lacuna with the network analysis method is that as the number of surfaces increases, it becomes increasingly messy to handle all the resistances. So in this book, we will not use the network analysis method but will instead use the radiosity irradiation method, which can be applied from a one surface enclosure to that with any number of surfaces.

4.5.1 Radiosity-Irradiation Method

As already mentioned, the credit for the development of this method goes to Professor
E.M.Sparrow and his colleagues at the University of Minnesota at Minneapolis. The
method is eminently programmable on the computer. It blends itself easily with
Computational Fluid Dynamics (CFD) calculations. In view of these for combined
heat transfer problems, this method is very potent. Before getting into the actual
method, we have to flesh out certain definitions.

Consider irradiation G_i falling on a surface that has a hemispherical total emissivity of ε_i which is maintained at a temperature T_i. A certain portion of the incident
radiation is reflected, given by $\rho_i G_i$ and the absorbed radiation is given by $\alpha_i G_i$.
Because the surface is at a temperature above 0 K, it also emits radiation given by
$\varepsilon_i \sigma T_i^4$ (Fig. 4.27).

Now, radiosity or the leaving flux or the leaving radiation $J_i (W/m^2)$ is given by

$$J_i = \varepsilon_i \sigma T_i^4 + \rho_i G_i \tag{4.179}$$

The incoming radiation is given by G_i. We are using the subscript i because the
expression can be for any surface in an enclosure.

The net radiation heat transfer from i, given by q_i is

$$q_i = J_i - G_i \tag{4.180}$$

For an opaque surface, transmission $\tau_i = 0$

$$\alpha_i + \rho_i + \tau_i = 1 \tag{4.181}$$

$$\therefore \rho_i = 1 - \alpha_i \tag{4.182}$$

Fig. 4.27 Depiction of
various radiation processes
on a surface

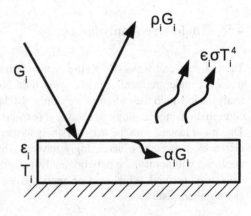

Substituting for ρ_i in Eq. (4.179), we get

$$J_i = \varepsilon_i \sigma T_i^4 + (1 - \alpha_i)G_i \qquad (4.183)$$

This is the radiosity relation for the ith surface of the enclosure. For a gray diffuse surface,

$$\alpha_i = \varepsilon_i \qquad (4.184)$$

$$\therefore \ J_i = \varepsilon_i \sigma T_i^4 + (1 - \varepsilon_i)G_i \qquad (4.185)$$

In Eq. 4.185, the first term represents the contribution from emission while the second is the contribution from reflection. It is easy to see that when $\varepsilon_i = 1$, $J_i = \sigma T_i^4$.

If we want to calculate the radiosity from a particular surface, we need to know the emissivity, the temperature and the irradiation falling on that surface. If the irradiation is because of radiation from several surfaces, we need to worry about the radiosity of these several surfaces. This is what the enclosure analysis is all about. But we are now in the preliminary stage and are trying to get an expression for the net radiation heat transfer so that we can use it eventually in the enclosure analysis.

For an ith surface of an enclosure,

$$q_i = J_i - G_i = \varepsilon_i \sigma T_i^4 + G_i - \varepsilon_i G_i - G_i \qquad (4.186)$$

$$q_i = \varepsilon_i(\sigma T_i^4 - G_i) \qquad (4.187)$$

So, if we are able to find just the irradiation on all the surfaces, we have solved the problem. However, it is not so easy to calculate the irradiation on a surface because the irradiation on a surface is because of the radiosity originating from other surfaces together with the geometric orientation accounted for. So, we must simultaneously solve either for the radiosities or for the irradiations. Working further on Eq. 4.186

$$q_i = J_i - G_i \qquad (4.188)$$

However, from Eq. 4.183 $G_i = \dfrac{J_i - \varepsilon_i \sigma T_i^4}{(1 - \varepsilon_i)} \qquad (4.189)$

Substituting for G_i in Eq. 4.188, we have

$$q_i = J_i - \frac{J_i - \varepsilon_i \sigma T_i^4}{(1 - \varepsilon_i)} \qquad (4.190)$$

$$q_i = \frac{J_i - \varepsilon_i J_i - J_i + \varepsilon_i \sigma T_i^4}{(1 - \varepsilon_i)} \qquad (4.191)$$

$$q_i = \frac{\varepsilon_i \left[\sigma T_i^4 - J_i\right]}{(1 - \varepsilon_i)} \qquad (4.192)$$

One may wonder what exactly is the difference between Eqs. (4.187) and (4.192) Both are expressions for heat flux and are correct! While Eq. (4.187) is in terms of irradiation, Eq. (4.192) is in terms of radiosity. Our original expression (Eq. 4.188) is in terms of both radiosity and irradiation. If we are evaluating the radiosity and irradiation for thousands of elements, there is no point in simultaneously storing both radiosity and irradiation. Information of radiosity can be obtained from irradiation and vice versa. Therefore, it makes sense to store only one of the two quantities. Some use just the irradiation method where they will solve for G_i's alone. Some use the radiosity method, where they solve for J_i's.

One word of caution. Equation 4.192 cannot be applied to black bodies as it has a singularity and the denominator becomes 0. When writing programs, we make ε_i for the black body as 0.995 or something close it and use it. Else we avoid Eq. 4.192 and instead use other expressions derived above for determining the net radiation heat transfer from such a surface.

4.5.2 Re-radiating Surface

A re-radiating surface is the radiation equivalent of an insulator. In an insulator, $q = 0$, where q can be q-conduction or q-convection. If q-radiation $= 0$, then it is called a re-radiating surface.

The question to ask now is "How do we engineer such a surface?" Take a surface and insulate it heavily on the back side so that no conduction heat transfer takes place and prevent all possible opportunities for heat transfer to take place. When we do all of these, whatever radiation is impinging on it must go out. For such a surface

$$J_i = G_i \tag{4.193}$$

$$q_i = 0 = \frac{\varepsilon_i \left[\sigma T_i^4 - J_i \right]}{(1 - \varepsilon_i)} \tag{4.194}$$

$$\therefore \; J_i = \sigma T_i^4 \tag{4.195}$$

It is a remarkable result because the radiosity of a re-radiating surface is independent of the emissivity and it will come to an equilibrium temperature of T_i depending on whatever irradiation it receives from the neighbouring bodies. This σT_i^4 is also equal to G_i. The temperature of the re-radiating surface is decided by its neighbours. Consider a three surface enclosure, where one surface is a hot surface, as in a heat treatment furnace, surface 2 houses the object to be heat treated and surface 3 is the intermediate surface which acts as the mediator. In this case, surface 3 takes the heat from the hot surface and passes it on to the other surfaces and will come to a temperature which is in between these two. Such re-radiating surfaces are frequently used in furnaces, combustion chambers and enclosures.

In passing, it is instructive to mention that for both a black surface and a re-radiating surface $J_i = \sigma T_i^4$. However, the latter has one more qualification. For a

Fig. 4.28 Enclosure under
consideration for the
formulation of the
radiosity-irradiation method

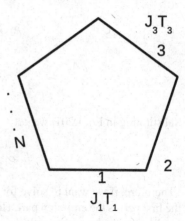

re-radiating surface $J_i = G_i$ so that $q_i = 0$. A simple, black surface need not satisfy
the above condition.

Consider an N surface enclosure as given in Fig. 4.28. Each of these surfaces is
characterized by a hemispherical spectral emissivity and temperature. All of these
are gray, diffuse surfaces. The total radiation leaving the ith surface is

$$A_i J_i = \varepsilon_i A_i \sigma T_i^4 + \rho_i \text{ (Incident)} \tag{4.196}$$

$$A_i J_i = \varepsilon_i A_i \sigma T_i^4 + (1 - \varepsilon_i) \text{ (Incident)} \tag{4.197}$$

What is this "Incident radiation" in the above two equations? It is intuitive that this
"Incident radiation" should be

"Incident"

$$= A_1 F_{11} J_1 + A_2 F_{21} J_2 + A_3 F_{31} J_3 + \cdots A_N F_{Ni} J_N = \sum_{j=1}^{N} A_j F_{ji} J_j \tag{4.198}$$

Substituting for "Incident" in Eq. 4.197

$$A_i J_i = A_i \varepsilon_i \sigma T_i^4 + (1 - \varepsilon_i) \sum_{j=1}^{N} A_j F_{ji} J_j \tag{4.199}$$

$$\text{But} \quad A_j F_{ji} = A_i F_{ij} \text{(reciprocal rule)} \tag{4.200}$$

$$\therefore \quad \cancel{A_i} J_i = \cancel{A_i} \varepsilon_i \sigma T_i^4 + \cancel{A_i} (1 - \varepsilon_i) \sum_{j=1}^{N} F_{ij} J_j \tag{4.201}$$

$$\text{However} A_i G_i = \sum_{j=1}^{N} A_j F_{ji} J_j \tag{4.202}$$

$$\cancel{A_i} G_i = \cancel{A_i} \sum_{j=1}^{N} F_{ij} J_j \tag{4.203}$$

$$\therefore G_i = \sum_{j=1}^{N} F_{ij} J_j \tag{4.204}$$

Substituting in Eq. 4.201, we get

$$J_i = \varepsilon_i \sigma T_i^4 + (1 - \varepsilon_i) G_i \tag{4.205}$$

Therefore, if we want to solve for the radiosity in an N surface enclosure problem, the first part is the emission part. Here, Stefan–Boltzmann's law is at work, which is $\varepsilon_i \sigma T^4$, for which we need to know the hemispherical spectral emissivity. We need to exploit Kirchhoff's law and the fact that the surface under consideration is an opaque, gray and diffuse surface, for which $\rho = 1 - \alpha$ and $1 - \alpha$ can be written as $1 - \varepsilon$. So our knowledge of radiative properties is being used here. When we write $F_{ij} J_j$, the view factors come into effect. So, all that we have studied thus far is incorporated in one single equation (Eq. 4.205), as seen above.

In an N surface enclosure, there are N such radiosity relations. If there are N equations and N unknowns, they can be easily solved simultaneously and we can get all the radiosities. Once we get all the radiosities, we can straightaway use the formula for the net radiative heat transfer from any surface in terms of radiosity. Otherwise, if it is only three surfaces, if we have the time and the patience, we can individually evaluate G_1, G_2 and G_3 for the 3 surfaces. We can then get $(J-G)$ for the say, three surfaces, and hence the heat fluxes.

Essentially, we have solved the problem of radiative heat transfer in an enclosure where there is no conduction and convection. If these two are present, we will write the additional energy equations and solve for them too. But what we have described above will continue to be the radiation portion of the solver. However, in our development, the enclosure has been assume dot be either evacuated or filled with a medium that does not participate in the radiation.

This is the enclosure theory developed by Prof. Sparrow and his colleagues. For up to 3 or maximum 4 surfaces, we can solve this using hand calculations. If the number of surfaces in the enclosure exceeds 4, we have to use the computer. Since it is a system of simultaneous equations, it is eminently solvable by the iterative Gauss–Seidel method. We do not have to invert. We usually start with some assumption like $J_1 = J_2 = J_3 = J_4 = 1000$ W/m^2 or a suitable number depending on the temperature range encountered in the problem.

Example 4.14 Consider a flat plate with an emissivity (ε), and maintained at a temperature T_i. It is placed in large surroundings at T_∞. The bottom of the plate is insulated. Using the radiosity-irradiation method, determine the net radiation heat transfer from the plate.

Fig. 4.29 A simple two surface enclosure problem

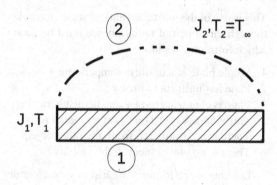

Solution:

First, we enclose the plate with an imaginary hemispherical basket (Fig. 4.29) at T_∞ for which $\varepsilon = 1$ and call this as surface 2. Now, we have an enclosure in hand.

View factor matrix for the two surface enclosures $= \begin{bmatrix} F_{11} & F_{12} \\ F_{21} & F_{22} \end{bmatrix}$

$$J_1 = \varepsilon_1 \sigma T_1^4 + (1 - \varepsilon_1)(F_{11} J_1 + F_{12} J_2) \tag{4.206}$$

$$J_2 = \varepsilon \sigma T_\infty^4 + (1 - \varepsilon_2)(F_{21} J_1 + F_{22} J_2) \tag{4.207}$$

$$J_2 = 1 \times \sigma T_\infty^4 + 0 \times (F_{21} J_1 + F_{22} J_2) \tag{4.208}$$

$$J_2 = \sigma T_\infty^4 \tag{4.209}$$

Now do we understand why we do not have to determine F_{21} and F_{22}?

Whenever we have surroundings like this where $\varepsilon = 1$, there is no need to waste time calculating those view factors.

$$J_1 = \varepsilon_1 \sigma T_1^4 + (1 - \varepsilon_1)(0 + 1 \times J_2) \tag{4.210}$$

$$J_1 = \varepsilon_1 \sigma T_1^4 + (1 - \varepsilon_1) J_2 \tag{4.211}$$

Substituting for J_2,

$$J_1 = \varepsilon_1 \sigma T_1^4 + (1 - \varepsilon_1)\sigma T_\infty^4 \tag{4.212}$$

$$G_1 = F_{11} J_1 + F_{12} J_2 = \sigma T_\infty^4 \tag{4.213}$$

$$q_1 = J_1 - G_1 \tag{4.214}$$

$$q_1 = \varepsilon_1 \sigma T_1^4 + \sigma T_\infty^4 - \varepsilon_1 \sigma T_\infty^4 - \sigma T_\infty^4 \tag{4.215}$$

$$q_1 = \varepsilon_1 \sigma (T_1^4 - T_\infty^4) \tag{4.216}$$

This is one of the most used and "abused" formula in heat transfer. Now, having gone through this much of radiation, we must be aware of the limiting conditions under which formula is valid.

1. Single plate at a uniform temperature
2. Plate has uniform radiosity
3. Plate is characterized by one hemispherical spectral emissivity
4. Surroundings are at constant temperature
5. There is no irradiation from any other object
6. There is no other object in the vicinity

Looking at the formula, it must now be clear under which situations this formula can be used and in which ones, its use is incorrect.

Example 4.15 Consider two parallel plates as shown in Fig. 4.30. They are infinitely deep in the top and bottom directions and extend infinitely in the direction perpendicular to the plane of the board. There is vacuum between the plates. The infinite extent is to basically help figure out that $F_{12} = 1$. Find out the net radiation heat transfer between the two surfaces using enclosure theory.

Fig. 4.30 Parallel plate geometry (Example 4.15)

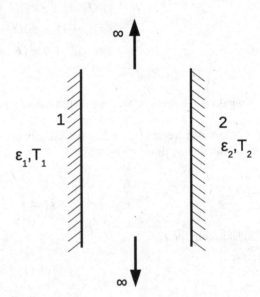

Solution:

$$\text{View factor matrix} = \begin{bmatrix} F_{11} & F_{12} \\ F_{21} & F_{22} \end{bmatrix} = \begin{bmatrix} 0 & 1 \\ 1 & 0 \end{bmatrix}$$

$$J_1 = \varepsilon_1 \sigma T_1^4 + (1 - \varepsilon_1) J_2 \tag{4.217}$$

$$J_2 = \varepsilon_2 \sigma T_2^4 + (1 - \varepsilon_2) J_1 \tag{4.218}$$

$$J_1 = \varepsilon_1 \sigma T_1^4 + (1 - \varepsilon_1)\left[\varepsilon_2 \sigma T_2^4 + (1 - \varepsilon_2)J_1\right] \quad (4.219)$$

$$(1 - (1 - \varepsilon_1)(1 - \varepsilon_2))J_1 = \varepsilon_1 \sigma T_1^4 + \varepsilon_2 \sigma T_2^4 - \varepsilon_1 \varepsilon_2 \sigma T_2^4 \quad (4.220)$$

$$J_1 = \frac{(\varepsilon_1 \sigma T_1^4 + \varepsilon_2 \sigma T_2^4 - \varepsilon_1 \varepsilon_2 \sigma T_2^4)}{[1 - (1 - \varepsilon_1)(1 - \varepsilon_2)]} \quad (4.221)$$

By inspection,

$$J_2 = \frac{(\varepsilon_1 \sigma T_1^4 + \varepsilon_2 \sigma T_2^4 - \varepsilon_1 \varepsilon_2 \sigma T_1^4)}{[1 - (1 - \varepsilon_1)(1 - \varepsilon_2)]} \quad (4.222)$$

$$q_1 = J_1 - G_1 \quad (4.223)$$

$$G_1 = F_{11} J_1 + F_{12} J_2 \quad (4.224)$$

$$G_1 = 0 + J_2 = J_2 \quad (4.225)$$

$$q_1 = J_1 - J_2 \quad (4.226)$$

$$q_1 = \frac{\varepsilon_1 \varepsilon_2 \sigma \left[T_1^4 - T_2^4\right]}{[1 - 1 + \varepsilon_1 + \varepsilon_2 - \varepsilon_1 \varepsilon_2]} \quad (4.227)$$

$$q_1 = \frac{\sigma \left[T_1^4 - T_2^4\right]}{\left[\frac{1}{\varepsilon_1} + \frac{1}{\varepsilon_2} - 1\right]} \quad (4.228)$$

This is called the **"Parallel plate formula"** which is a very powerful formula in radiation. Let us now look at the asymptotic correctness of the result. What we mean by this is that when we apply the result to an extreme case, it should work. Let us check this out for this problem. Suppose surface 2 were to be the surroundings at T_∞, then $\varepsilon_2 = 1$ and $\frac{1}{\varepsilon_2} = 1$, $T_2 = T_\infty$. Hence, the formula for q would reduce to $q = \varepsilon_1 \sigma(T_1^4 - T_\infty^4)$ which is the result for net radiation heat transfer from a single surface, a formula we derived a little while ago. Therefore, for the asymptotic limit of the second surface being the ambient, this formula works.

If we have 2 parallel plates and even though we do not have convection between them, if the temperature difference between the 2 plates is sufficiently large, high radiation heat transfer between them is inevitable. But many times we want to avoid this radiation heat transfer between them and just because there is vacuum between the plates, it does not mean that we have solved the problem. If both the surfaces have good emissivity and have a good temperature difference, the radiation heat transfer will not be insignificant. Therefore, the challenge is now to come up with some method by which we reduce the radiation heat transfer between the 2 surfaces. One possibility is to insert a thin film, and we try to find out what the heat transfer will be when such a shield is inserted between the 2 plates. Once we derive the heat transfer with one such shield, by induction, we can find out what the heat transfer will be if there are 2 shields, 3 shields or n such shields.

Example 4.16 Consider two infinitely long parallel plates that are at temperatures T_1 and T_2, respectively, with hemispherical total emissivities ε_1 and ε_2. The inter-

Fig. 4.31 Radiation
between parallel plates in the
presence of a shield

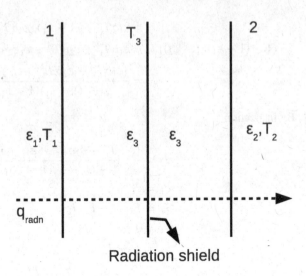

Radiation shield

vening space is evacuated. The radiative heat transfer has already been derived for
such a case. Now, we insert a radiation shield that has emissivity ε_3 on both sides
(Fig. 4.31). We have to say "both sides" because emissivity is a surface property and
by engineering different coatings on both sides, we can have two different emissivities too. The whole shield may be at one temperature but the two sides may have
different emissivities. Let the temperature of the shield be T_3. Steady state prevails
in the system. Radiation heat transfer is taking place as shown. Come up with a
mathematical expression for q with the shield.

Solution:

With the shield

$$q_{13} = \frac{\sigma\left[T_1^4 - T_3^4\right]}{\left[\frac{1}{\varepsilon_1} + \frac{1}{\varepsilon_3} - 1\right]} \tag{4.229}$$

$$q_{32} = \frac{\sigma\left[T_3^4 - T_2^4\right]}{\left[\frac{1}{\varepsilon_3} + \frac{1}{\varepsilon_2} - 1\right]} \tag{4.230}$$

Under steady state, $q_{13} = q_{32} = q_{12,\text{shield}}$

$$q_{12,\text{shield}} = \frac{\sigma\left[T_1^4 - T_3^4\right]}{\left[\frac{1}{\varepsilon_1} + \frac{1}{\varepsilon_3} - 1\right]} \tag{4.231}$$

$$= \frac{\sigma\left[T_3^4 - T_2^4\right]}{\left[\frac{1}{\varepsilon_3} + \frac{1}{\varepsilon_2} - 1\right]} \tag{4.232}$$

We do not know T_3 at this point in time as it is the equilibrium temperature. Instead of, say, inserting thermocouples and determining T_3, let us try and eliminate T_3. If $\frac{1}{2} = \frac{2}{4}$ then this is also equal to $\frac{1+2}{2+4} = \frac{3}{6} = \frac{1}{2}$ (Dividendo componendo rule). Using this rule,

$$q_{12,\text{shield}} = \frac{\sigma\left[T_1^4 - \cancel{T_3^4} + \cancel{T_3^4} - T_2^4\right]}{\left[\frac{1}{\varepsilon_1} + \frac{1}{\varepsilon_2} + \frac{2}{\varepsilon_3} - 2\right]} \tag{4.233}$$

$$q_{12,\text{shield}} = \frac{\sigma\left[T_1^4 - T_2^4\right]}{\left[\frac{1}{\varepsilon_1} + \frac{1}{\varepsilon_2} + \frac{2}{\varepsilon_3} - 2\right]} \tag{4.234}$$

If $\varepsilon_1 = \varepsilon_2 = \varepsilon_3 = \varepsilon$,   then

$$q_{12,\text{shield}} = \frac{\sigma\left[T_1^4 - T_2^4\right]}{2\left[\frac{2}{\varepsilon} - 1\right]} \tag{4.235}$$

If there were no shield and if $\varepsilon_1 = \varepsilon_2 = \varepsilon$, then from Eq. 4.228

$$q_{12,\text{noshield}} = \frac{\sigma[T_1^4 - T_2^4]}{[\frac{2}{\varepsilon} - 1]} \tag{4.236}$$

$$\therefore q_{12,\text{shield}} = \frac{q_{12,\text{noshield}}}{2} \tag{4.237}$$

If n such shields are inserted, it is intuitively clear that

$$q_{12,\text{shield}} = \frac{q_{12,\text{noshield}}}{n+1} \tag{4.238}$$

Therefore, it is possible to insulate surfaces radiatively by employing "n" number of shields. Just because we have vacuum, it does not mean that we have insulated the plates, it is just that we have removed the convection. Radiation still will be present. But if we have evacuated the surfaces and have n shields placed in between, the radiation can be substantially reduced. Sometimes this is also referred to as a super insulation. It is apparent that the position of the sheet does not actually matter.

In the above example, while the heat transfer rate is easy to determine, if we want to know if the radiation sheet that is introduced can withstand some temperature, it is imperative to evaluate T_3. This has to be evaluated as a post-processed quantity, and we use this information to see whether it is within limits of the material of the shield.

Example 4.17 Determine the steady state temperatures of two radiation shields placed in the evacuated space between two infinite plates at temperatures 600 K and 300 K, respectively. All the surfaces are gray and diffuse with emissivities of 0.85 (Fig. 4.32).

Fig. 4.32 Problem geometry
(Example 4.17)

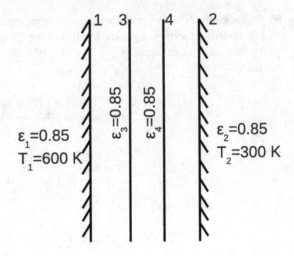

Solution:

$$q_{12,\text{noshield}} = \frac{\sigma\left[T_1^4 - T_2^4\right]}{\left[\frac{2}{\varepsilon} - 1\right]} \tag{4.239}$$

$$= \frac{5.67 \times 10^{-8}\left[600^4 - 300^4\right]}{\left[\frac{2}{0.85} - 1\right]} \tag{4.240}$$

$$= 5091.9 \text{ W/m}^2$$

$$q_{12,\text{2shield}} = \frac{q_{12,\text{noshield}}}{2+1} = \frac{5091.6}{3} \tag{4.241}$$

$$= 1697.3 \text{ W/m}^2$$

$T_3 = ?$

$$\frac{\sigma\left[T_1^4 - T_3^4\right]}{\left[\frac{2}{\varepsilon} - 1\right]} = 1697.3 \tag{4.242}$$

$$\frac{5.67 \times 10^{-8}\left[600^4 - T_3^4\right]}{\left[\frac{2}{0.85} - 1\right]} = 1697.3 \tag{4.243}$$

$$T_3 = 587.5 \text{ K} \tag{4.244}$$

$T_4 = ?$

$$\frac{\sigma\left[T_3^4 - T_4^4\right]}{\left[\frac{2}{\varepsilon} - 1\right]} = 1697.3 \tag{4.245}$$

$$\frac{5.67 \times 10^{-8} \left[587.5^4 - T_4^4\right]}{\left[\frac{2}{0.85} - 1\right]} = 1697.3 \qquad (4.246)$$

$$T_4 = 529.5 \text{ K} \qquad (4.247)$$

Example 4.18 A hole 5 mm in diameter and 25 mm deep is bored in a gray, diffuse material (Fig. 4.33). The emissivity of all the surfaces is 0.6 and is maintained at a uniform temperature of 1000 K. The surroundings are at 300 K. Determine the net radiant heat transfer leaving the opening of the cavity.

Fig. 4.33 Problem geometry (Example 4.18)

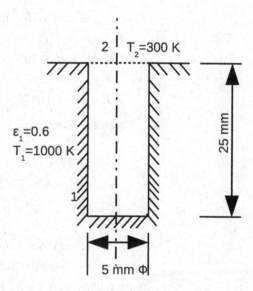

Explanation:

We have a flat bottomed hole, which has been bored into a plate. Its depth is 25 mm and its diameter is 5 mm. The hole is evacuated or filled with air which is radiatively non-participating. The temperature of surface 1, which includes the bottom and the lateral surface area of the hole is 1000 K. Its emissivity $\varepsilon_1 = 0.6$. The bored hole is opening to the surroundings at 300 K. We can now treat it as a 2 zone enclosure. The beauty of the problem is that, suppose the hole was not there and we have surface 2 alone, it would be a circular disc of diameter 5mm. But now we say that this circle is a black body, which is at 1000 K. This will dissipate certain amount of heat to the surroundings at 300 K. We want to see, compared to this, how much the bored hole will dissipate? The ratio of these two is called **effective emissivity**. As the depth of the bored hole increases, its effective emissivity will approach 1. Suppose we do not get a surface with a good emissivity, it is possible for us to bore holes at a few places and augment the heat transfer passively, without using any pumping power.

Solution:

(a) View factor $F_{21} = 1$; $F_{22} = 0$

$$A_1 = 2\pi rh + \pi r^2 \tag{4.248}$$
$$A_1 = (2\pi \times 0.0025 \times 0.025) + \pi \times 0.0025^2 \tag{4.249}$$
$$A_1 = 4.121 \times 10^{-4} \text{ m}^2$$
$$A_2 = \pi r^2 = \pi \times 0.0025^2 \tag{4.250}$$
$$A_2 = 1.96 \times 10^{-5} \text{ m}^2$$
$$A_1 F_{12} = A_2 F_{21} \tag{4.251}$$
$$4.121 \times 10^{-4} \times F_{12} = 1.96 \times 10^{-5} \times 1 \tag{4.252}$$
$$F_{12} = 0.0475$$
$$F_{11} = 1 - 0.0475 = 0.9525$$
$$\text{View factor matrix} = \begin{bmatrix} 0.9525 & 0.0475 \\ 1 & 0 \end{bmatrix}$$

$$J_1 = \varepsilon_1 \sigma T_1^4 + (1 - \varepsilon_1)[F_{11}J_1 + F_{12}J_2] \tag{4.253}$$
$$J_2 = \sigma T_\infty^4 = 459.3 \text{ W/m}^2 \tag{4.254}$$
$$q_1 = \frac{\varepsilon_1}{1 - \varepsilon_1}\left[\sigma T_1^4 - J_1\right] \tag{4.255}$$

$$J_1 = 0.6 \times 5.67 \times 10^{-8} \times (1000)^4$$
$$+ 0.4[0.9525 J_1 + 0.0475 \times 459.3] \tag{4.256}$$
$$J_1[1 - 0.4 \times 0.9525] = 0.6 \times 5.67 \times 10^{-8} \times (1000)^4$$
$$+ 0.4 \times 0.0475 \times 459.3 \tag{4.257}$$
$$0.6188 J_1 = 34020 + 8.73$$
$$\therefore J_1 = 54991.5 \text{ W/m}^2 \tag{4.258}$$
$$q_1 = \frac{0.6}{0.4}[56700 - 54991.5]$$
$$q_1 = 2562.8 \text{ W/m}^2 \tag{4.259}$$
$$Q_1 = q_1 \times A_1 = 1.06 \text{ W} \tag{4.260}$$

(b)

$$Q_{1(\text{blackbody})} = A_2 \times \sigma\left[T_1^4 - 300^4\right]$$
$$= 1.102 \text{ W} \tag{4.261}$$
$$\varepsilon_{\text{eff}} = \frac{1.06}{1.102} = 0.961 \tag{4.262}$$

Hence, the effective emissivity of the flat bottomed hole is 0.96 which is very close to that of a black body. What this means is that if we just had a black circular disk 5 mm in diameter at 1000 K, radiating to the surroundings at 300 K, the radiant energy transferred would have been 1.1 W. As opposed to this, with the hole we obtain a value close to 1.06 W. Please note that the original surface had an emissivity of only 0.6 and if we did not have the hole, the disk would have radiated 0.661 W. Hence, there is an augmentation in the heat transfer which will decrease as $\varepsilon \rightarrow 1$ (of the parent surface)

Example 4.19 A very long electrical conductor 10mm diameter is concentric with a cooled cylindrical tube 50 mm in diameter whose surface is gray and diffuse with an emissivity of 0.9 and a temperature of 300 K (Fig. 4.34). The electrical conductor has a diffuse gray surface with an emissivity of 0.6 and dissipates 7 W/m length. Assuming that the space between the conductor and the tube is evacuated, determine the surface temperature of the conductor.

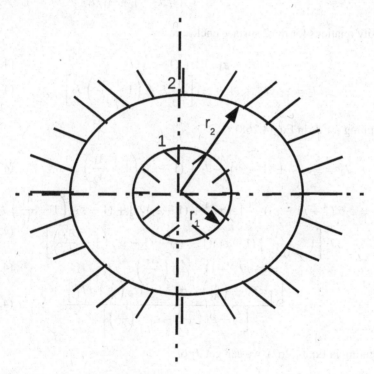

Fig. 4.34 Problem geometry (Example 4.19)

Solution:

We can say that this is an inverse problem where we know the heat flux but do not know the temperature. We need to find the temperature to see if the wire can withstand it or not. Inverse problems are more practical from an engineer's perspective than a direct problem. Many a time, we have to infer, as we cannot straightaway solve this without first writing the radiosity relations and getting the formula. It is better that we name the two surfaces as surface 1 and surface 2 with areas A_1 and A_2, start with view factors and get expressions for J_1 and J_2. Next, we get a general expression for Q_1 and then substitute the specific values to get the answer we are seeking. We call these areas A_1 and A_2. If a problem of sphere within a sphere is encountered, the area formulae alone need be substituted in that case.

$$F_{11} = 0, F_{12} = 1 \tag{4.263}$$
$$A_1 F_{12} = A_2 F_{21} \tag{4.264}$$
$$F_{21} = A_1/A_2 \tag{4.265}$$
$$F_{22} = 1 - A_1/A_2 \tag{4.266}$$
$$\text{View factor matrix} = \begin{bmatrix} 0 & 1 \\ A_1/A_2 & 1 - A_1/A_2 \end{bmatrix}$$

Radiosity relations for the 2 surface enclosure

$$J_1 = \varepsilon_1 \sigma T_1^4 + (1 - \varepsilon_1) J_2 \tag{4.267}$$
$$J_2 = \varepsilon_2 \sigma T_2^4 + (1 - \varepsilon_2)\left[\frac{A_1}{A_2} J_1 + \left(1 - \frac{A_1}{A_2}\right) J_2\right] \tag{4.268}$$

Substituting for J_1 in Eq. (4.268)

$$J_2 = \varepsilon_2 \sigma T_2^4 + (1 - \varepsilon_2)\frac{A_1}{A_2} J_1 + (1 - \varepsilon_2)\left(1 - \frac{A_1}{A_2}\right) J_2 \tag{4.269}$$

$$J_2 = \varepsilon_2 \sigma T_2^4 + (1 - \varepsilon_2)\frac{A_1}{A_2}\left[\varepsilon_1 \sigma T_1^4 + (1 - \varepsilon_1)J_2\right] + (1 - \varepsilon_2)\left(1 - \frac{A_1}{A_2}\right) J_2$$
$$J_2\left[1 - \left(\frac{A_1}{A_2}\right)(1 - \varepsilon_1)(1 - \varepsilon_2) - (1 - \varepsilon_2)\left(1 - \frac{A_1}{A_2}\right)\right] \tag{4.270}$$
$$= \varepsilon_2 \sigma T_2^4 + (1 - \varepsilon_2)\left(\frac{A_1}{A_2}\right)\varepsilon_1 \sigma T_1^4 \tag{4.271}$$
$$J_2 = \frac{\left[\varepsilon_2 \sigma T_2^4 + \left(\frac{A_1}{A_2}\right)\varepsilon_1 \sigma T_1^4 - \varepsilon_1\varepsilon_2\left(\frac{A_1}{A_2}\right)\sigma T_1^4\right]}{\left[\varepsilon_2 + \varepsilon_1\left(\frac{A_1}{A_2}\right) - \varepsilon_1\varepsilon_2\left(\frac{A_1}{A_2}\right)\right]} \tag{4.272}$$

Substituting in Eq. (4.267), we can get J_1 or

$$q_{12} = J_1 - J_2 \tag{4.273}$$
$$q_{12} = \varepsilon_1 \sigma T_1^4 + (1 - \varepsilon_1)J_2 - J_2 \tag{4.274}$$

$$q_{12} = \varepsilon_1 \sigma T_1^4 + \cancel{J_2} - \varepsilon_1)J_2 - \cancel{J_2} \tag{4.275}$$

$$q_{12} = \varepsilon_1 \sigma T_1^4 - \varepsilon_1 J_2 \tag{4.276}$$

$$q_{12} = \varepsilon_1 \sigma T_1^4 - \varepsilon_1 \left[\frac{\left[\varepsilon_2 \sigma T_2^4 + \left(\frac{A_1}{A_2}\right) \varepsilon_1 \sigma T_1^4 - \varepsilon_1 \varepsilon_2 \left(\frac{A_1}{A_2}\right) \sigma T_1^4 \right]}{\left[\varepsilon_2 + \varepsilon_1 \left(\frac{A_1}{A_2}\right) - \varepsilon_1 \varepsilon_2 \left(\frac{A_1}{A_2}\right) \right]} \right] \tag{4.277}$$

$$q_{12} = \frac{\varepsilon_1 \varepsilon_2 \sigma [T_1^4 - T_2^4]}{\left[\varepsilon_2 + \varepsilon_1 \left(\frac{A_1}{A_2}\right) - \varepsilon_1 \varepsilon_2 \left(\frac{A_1}{A_2}\right) \right]} \tag{4.278}$$

$$q_{12} = \frac{\sigma [T_1^4 - T_2^4]}{\left[\frac{1}{\varepsilon_1} + \frac{1}{\varepsilon_2} \left(\frac{A_1}{A_2}\right) - \left(\frac{A_1}{A_2}\right) \right]} \tag{4.279}$$

$$Q_{12} = q_{12} 2\pi r_1 L$$

A good check for this formula is that if $A_1/A_2 = 1$, Eq. 4.279 will reduce to the parallel plate formula. Equation 4.279 is valid for two concentric spheres, as well. So, this formula displays asymptotic correctness. Substituting for the values of $\varepsilon_1, \varepsilon_2, A_1, A_2$ and T_2 in Eq. 4.279,

$$\frac{A_1}{A_2} = \frac{2\pi r_1 \cancel{L}}{2\pi r_2 \cancel{L}} = \frac{r_1}{r_2} = 0.2 \tag{4.280}$$

$$\frac{7}{2\pi r_1 (1)} = \frac{6.7 \times 10^{-8}[T_1^4 - 300^4]}{\frac{1}{0.6} + \frac{1}{0.9}(0.2) - 0.2}$$

$$T_1 = 348.4 \text{ K} \tag{4.281}$$

The wire can withstand this temperature, roughly 75 °C. Whenever the view factors involve the fundamental dimensions such as r_1 and r_2 in this case, the analysis becomes complicated but towards the end, we get a simplified answer.

Example 4.20 Consider a very deep triangular duct (deep in the direction perpendicular to the plane of the board) made of diffuse gray walls, each of which has a width of 1.5 m (Fig. 4.35). The temperatures of surfaces 1 and 2 are 1200 K and 800 K respectively. The corresponding emissivities are 0.4 and 0.6, respectively. Surface 3 is completely insulated and has an emissivity of 0.5. For this two-dimensional enclosure,

a. Determine the net radiation heat transfer from surface 1.
b. Determine the temperature of the insulated surface 3.
c. If ε_3 is changed, will your results change?

Fig. 4.35 Problem geometry
(Example 4.20)

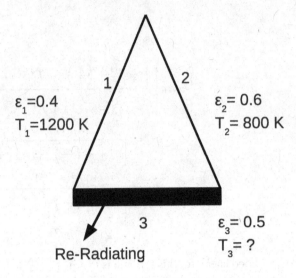

Re-Radiating

Solution:
a.

$$\text{View factor matrix} = \begin{bmatrix} 0 & 0.5 & 0.5 \\ 0.5 & 0 & 0.5 \\ 0.5 & 0.5 & 0 \end{bmatrix}$$

$$J_1 = \varepsilon_1 \sigma T_1^4 + (1 - \varepsilon_1)[0.5 J_2 + 0.5 J_3] \tag{4.282}$$

$$J_2 = \varepsilon_2 \sigma T_2^4 + (1 - \varepsilon_2)[0.5 J_1 + 0.5 J_3] \tag{4.283}$$

$$J_3 = G_3 = F_1 J_1 + F_{32} J_2 + F_{33} J_3 = 0.5[J_1 + J_2] \tag{4.284}$$

This is a crucial step in enclosure with re-radiating surface. Even if we have N surfaces, we have to solve for $N-1$ equations only, because for the re-radiating surface, the radiosities can be directly expressed in terms of the radiosities of the other surfaces. So, instead of solving 3 simultaneous equations, we need to solve for only 2.

Substituting the values for the variables, we get

$$J_1 = 0.4 \times 5.67 \times 10^{-8} \times 1200^4 + 0.6[0.5 J_2 + 0.5 J_3] \tag{4.285}$$

$$J_1 = 47029.3 + 0.3 J_2 + 0.3 J_3 \tag{4.286}$$

$$J_2 = 13934.6 + 0.2 J_1 + 0.2 J_3 \tag{4.287}$$

We substitute for J_3 in terms of J_1 and J_2, which reduces the 2 Equations 4.286 and 4.287 having only variables J_1 and J_2

$$J_1 = 47029.3 + 0.3 J_2 + 0.3[0.5 J_1 + 0.5 J_2] \tag{4.288}$$

$$J_2 = 13934.6 + 0.2 J_1 + 0.2[0.5 J_1 + 0.5 J_2] \tag{4.289}$$

$$0.85 J_1 = 47029.3 + 0.45 J_2 \qquad (4.290)$$
$$0.9 J_2 = 13934.6 + 0.3 J_1 \qquad (4.291)$$

Equations (4.290) and (4.291) can be solved to get J_1 and J_2

$$J_1 = 77137.9 \ \text{W/m}^2$$
$$J_2 = 41195.5 \ \text{W/m}^2$$
$$J_3 = \frac{J_1 + J_2}{2} = 59166 \ \text{W/m}^2$$

$$q_1 = \frac{\varepsilon_1}{1 - \varepsilon_1}[\sigma T_1^4 - J_1] \qquad (4.292)$$
$$q_1 = 2.69 \times 10^9 \ \text{W/m}^2 \qquad (4.293)$$
$$Q_1 = q_1 A_1 = 2.69 \times 10^4 \times 1.5 \times 1$$
$$= 4.04 \times 10^9 \ \text{W/m} \qquad (4.294)$$
$$J_3 = \sigma T_3^4 \qquad (4.295)$$

b.

$$T_3 = \left(\frac{J_3}{\sigma}\right)^{0.25} \qquad (4.296)$$
$$= 1010.7 \ \text{K}$$

Surface 3 is just receiving heat from surface 1 and is transferring this onto surface 2. So, it has no net radiation heat transfer. What is going out is equal to what is coming in, which is why it is a re-radiating surface.

c. If the emissivity is changed, it has no bearing. So the specification, emissivity = 0.5 for surface 3 is superfluous or redundant and is never used in the calculations. The good thing about a re-radiating surface is that we never worry about its emissivity. The temperature T_3 is actually called the equilibrium temperature of the re-radiating surface.

Problems

4.1 In a rectangular box type enclosure consisting of 6 surfaces, how many factors are there in total? How many independent view factors need to be determined?

4.2 A very long duct has the shape of a regular pentagon. How many view factors need to be independently evaluated? Determine the view factor between any two adjacent sides.

4.3 A long duct has a regular hexagonal cross-section. Determine the view factor between the opposite sides by view factor algebra. Cross check the answer

Fig. 4.36 Figure for Problem 4.4

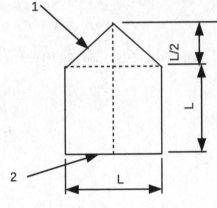

Fig. 4.37 Geometry for Problem 4.6

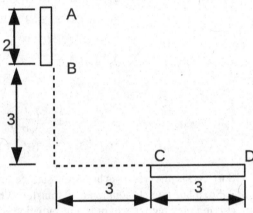

with the direct application of crossed string method to determine the same view factor.

4.4 Consider a glass house like structure shown in Fig. 4.36. All the walls are infinitely deep in the direction perpendicular to the plane of the paper. Evaluate the view factor F_{12} by any method known to you (No integration). Please note that "2" is the full bottom inside surface.

4.5 Consider a rectangular box type enclosure. This enclosure (radiation from the outer surfaces of the six walls is not part of the analysis) is cut vertically at the middle by an imaginary wall so that the top and bottom surfaces are divided into two surfaces each. The two surfaces now constituting the top are 1 and 4 while the two at the bottom are 3 and 4 with 3 placed right below 1. Each area is denoted using the letter A followed by the subscript. For example, the area of surface is A_1. Show that for this enclosure, the view factor F_{12} is given by

$$F_{12} = \frac{1}{2A_1}[A_{14}F_{14-23} - A_1 F_{13} - A_4 F_{42}]$$

4.6 Consider two rectangular thin strips AB and CD with dimensions and orientation as seen in Fig. 4.37 (all dimensions are in m). The strips are infinitely deep in the direction perpendicular to the plane of the paper. Determine the view factor F_{AB-CD} by using

    (a) Hottel's crossed string method
    (b) The decomposition rule (together with the simple view factor formula for a triangular enclosure)

4.7 Consider two semi-circular surfaces of radius R, separated by a minimum distance of S, as shown in Fig. 4.38 The two surfaces are infinitely deep in the plane perpendicular to the plane of the paper. Determine the view factor F_{12} by an intelligent application of Hottel's crossed string method (Hint: We may have to use arc lengths in the Hottel's method).

4.8 Consider a vertical enclosure in the shape of the frustum of a cone with the bottom surface having a diameter of 3.5 m and the top surface having a diameter of 1.75 m. The height of the enclosure is 5 m. Using view factor charts and algebra, treating the bottom surface as 1, top as 2 and the lateral surface as 3, determine all the view factors for the three surface enclosure.

4.9 A cubical furnace is 1.5 m long on all the sides. Evaluate all the view factors.

4.10 Consider the situation of a clear night sky in a desert. The minimum sky temperature on a particular night is −42 °C. Determine the temperature on the surface of a shallow pond of water if the ambient temperature is 24 °C and the natural convection coefficient for air is 5.5 W/m^{2}K. Take the emissivity of water to be 0.98 (Clue: Write down the energy balance equation and solve it iteratively for the temperature of the water).

Fig. 4.38 Geometry for Problem 4.7

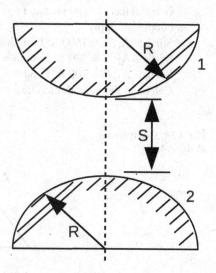

4.11 Consider two very long, gray, diffuse parallel plates that are separated by a small distance. The space between the two plates is evacuated. The left plate is at 700 K and the right plate is at 400 K. Two radiation shields, which are also gray and diffuse, are placed between the two plates. All surfaces have an emissivity of 0.8. Determine the steady state temperature of the two shields.

4.12 A spherical tank of diameter $D_1 = 0.52$ m containing liquid nitrogen is enclosed inside another spherical tank of diameter $D_2 = 0.81$ m and the space between them is evacuated. The inner and outer spheres are maintained at $T_1 = 80$ K and $T_2 = 270$ K, respectively. Both spheres have an emissivity of 0.07. Calculate the rate of transfer to the inner sphere and the rate of evaporation, if the latent heat of vaporization of liquid nitrogen is 2×10^5 Ws/kg.

4.13 The annular space between two concentric tubes having diameters of 20 and 50 mm is evacuated. The outer surface of the inner tube, which is diffuse and gray with an emissivity of 0.02, is maintained at a temperature of 255 K. The inner surface of the larger tube, with an emissivity of 0.05, is maintained at a temperature of 303 K. Determine the radiative heat transfer in the annular space between the tubes by treating this as a two surface enclosure. If a thin-walled radiation shield, that is diffuse and gray with an emissivity of 0.02 (both sides) is inserted in the middle (i.e. between the inner and outer surfaces), calculate the heat transfer rate in the presence of the shield.

4.14 Consider a cylindrical evacuated enclosure with height 1 m and radius 0.4 m. The top wall is maintained at 400 K and has an emissivity of 0.8, while the bottom wall is at 800 K and is black. The lateral wall of the enclosure is maintained at 600 K and has an emissivity of 0.4. Evaluate the net radiative heat transfer from all the three surfaces (not just the heat flux alone!) and thereby verify the energy balance.

4.15 A two-dimensional gray diffuse evacuated enclosure (with no heat transfer to outside) has each surface at a uniform temperature (Fig. 4.39). The following conditions apply
Surface A_1: T = 1500 K, Length = 4 m and $\epsilon = 0.6$
Surface A_2: T = 300 K, Length = 3 m and $\epsilon = 0.9$
Surface A_3: T = 700 K, Length = 4.5 m and $\epsilon = 0.5$
Calculate the net radiation heat transfer from all the three surfaces.

Fig. 4.39 Geometry for
Problem 4.15

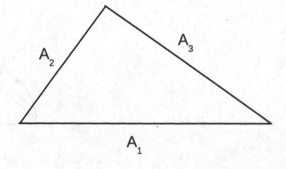

4.16 An evacuated, gray diffuse enclosure is in the shape of a triangle ABC. The
sides of the enclosure are infinitely deep in the direction perpendicular to the
plane of the paper. Details of the geometry and the properties of the three
surfaces are given in Table 4.1

  (a) Obtain all the view factors using view factor algebra.
  (b) Formulate the problem using the radiosity-irradiation method.
  (c) Solve for the radiosities and determine both the fluxes and the heat transfer
     rates at all the three walls.
     Properties of the triangular enclosure surfaces are listed in Table 4.1.

Table 4.1 Details of the geometry and the properties of the surfaces of the triangular enclosure
under consideration in Problem 4.16

Surface	Length, m	Emissivity, ϵ	Temperature, K
AB	7.1	0.9	1200
BC	6.2	0.6	400
CA	5.3	0.1	Reradiating

A_3, T_3=600 K, ε_3=0.6

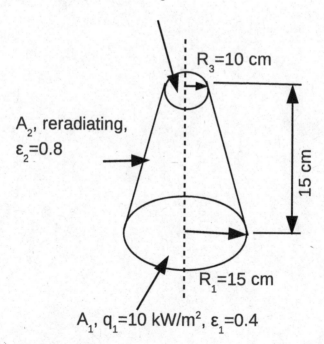

R_3=10 cm

A_2, reradiating,
ε_2=0.8

15 cm

R_1=15 cm

A_1, q_1=10 kW/m², ε_1=0.4

Fig. 4.40 Geometry for Problem 4.17

4.17 A frustum of a cone has its base heated as shown in Fig. 4.40. The top is held at 600 K while the side is perfectly insulated. All surfaces are diffuse gray. The pertinent dimensions, properties and heat flux from the surface 1 are shown in figure. Treating this is a three surface enclosure problem, determine the temperature attained by surface 1 as a result of radiative exchange within the enclosure?

Chapter 5
Radiation in Participating Media

A participating medium is an absorbing, emitting and scattering medium.

- Any particle at a temperature more than 0 K will emit radiation. Absorption is different, in the sense that, if 100 W/m² radiation is incident on a gas volume, what comes out will be less than that, as it absorbs a certain portion of the electromagnetic radiation.
- What such a medium can do further is to receive the radiation and reflect it in several directions, which is called scattering. From a volume, there can be out-scattering (the name given to the radiation going out) and in-scattering too. Out-scattering need not be the same in all the directions, which means that the medium is anisotropic. How much a medium scatters can be a function of θ. Treatment of scattering can thus get very complex!

The study of heat transfer through media which can absorb, emit and scatter radiation has been receiving increasing attention in the last few decades. In fact, it gained a lot of momentum after Nobel laureate Prof. Subramaniam Chandrashekar among others made seminal contributions to the equation of radiative heat transfer and its solution. The interest in this field arises from phenomena associated with rocket propulsion, combustion chambers, ablating systems, nuclear fusion and insulating systems. In all these, we have gases which are participating, unlike transparent media like air.

For example, in a class room which contains air, the radiation from the left wall will directly go to the right wall. Air does not participate in this; it just is at some temperature and convection may take place. The moment we have carbon dioxide or water vapour in large quantities inside the room, the air will absorb and or scatter. The absorption and emission of these gases in general are spectrally dependent and these also vary with the temperature. If we look at the absorption of solar radiation by the earth, we will see that the radiation is incoming from a black body at 6000K, while the outgoing radiation is from a body at 288 or 300 K. The incoming is mostly in the visible region while the outgoing is mostly in the infrared region and if the gases which are in the atmosphere are such that they permit the incoming energy and

C. Balaji, *Essentials of Radiation Heat Transfer*,
https://doi.org/10.1007/978-3-030-62617-4_5

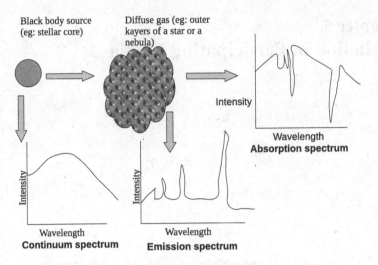

Fig. 5.1 Stellar structure and emission/absorption spectra

do not permit the outgoing energy, what happens is that there is a constant buildup of energy within the earth's atmosphere.

Some of the above phenomena are new while others are hundreds years old. Man was always interested in atmospheric radiation. Astrophysicists have been interested in gas radiation with regard to studying the structure of stars and radiation coming out of them. The spectrum observed during emission or absorption of radiation by a gas is characteristic of that gas alone. From the spectrum, we can get the signature, which is diagnostic of the gas present in the star. If this is studied over varied periods of time, we can find out if the gas concentration is changing or the star is moving towards or away from us and so on. Hence, the spectrum can be used as a diagnostic tool to determine the gas temperature, concentration, a star's speed and so on.

If we consider a black body source or outer layers of a star structure, we can look at two spectra, which are the emission spectra and the absorption spectra. The signature is obtained as intensity versus wavelength. From this, we can figure out the gas concentration and the gas temperature. This is basically an inverse problem, as from the output, we have to guess the input. There can be various causes for such a spectrum, and the goal of a successful inverse methodology is to correctly identify the cause which led to such a behaviour.

A simplified representation of the same is given in Fig. 5.1. We have a hot source and then a gas. We can get a continuous spectrum or an absorption spectrum or an emission spectrum in the form of emission bands, as seen here.

If we look at the incident solar energy flux that is coming on to the earth, as shown in Fig. 5.2, we can see that the first dashed line gives the energy distribution of a black body at 6000 K. The second curve shows the actual solar irradiance outside the atmosphere. The inner curve shows the solar irradiance after passing

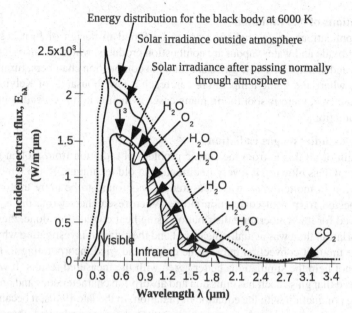

Fig. 5.2 Attenuation of incident solar spectral energy flux by the earth's atmosphere

normally through the atmosphere, which we see has a jagged shape, because there are absorption bands due to oxygen, carbon dioxide and water vapour molecules.

We can determine the absorption by having an instrument that has a sensor, which will exactly capture this between two particular wavelengths. We can see that ozone absorption is maximum in the visible part of the spectrum while carbon dioxide presence is felt more in the infrared region. Water vapour absorption is present almost throughout and hence we can design an instrument to measure it by having multiple channels. This instrument can be housed in a geostationary satellite, and hence will be called a **multispectral instrument or sounder**.

In passive remote sensing, the radiation emitted from the earth's surface that is absorbed and scattered by the atmospheric constituents is usually measured by a satellite. This is done through radiometric sensors placed at the top of the polar or geostationary satellite and is called **top of the atmosphere radiance**. The intensity itself is a matrix consisting of spectral intensities with two polarizations each, which is the spectral signature. The inverse problem of radiative transfer is concerned with estimating the atmospheric constituents from the spectral signature.

What is the logic on which this works? We will assume some atmospheric concentration, solve pertinent equations and determine q_λ at various λs. Then, we try to match our prediction with what is measured. The two will not agree in general and hence we iterate. We keep doing this till the measurements and simulations match closely. This guessing is easier said than done. What was described so far as one word "guessing" is the field of "inverse problems"!

Applications of gas radiation

Other applications of gas radiation are to be found in design of furnaces where carbon dioxide and water vapour are combustion products, which are also significant emitters and absorbers. These gases are found in combustion chambers, furnaces, IC engines, where the flame temperature can reach a few thousands of Kelvin. Apart from these two, there is soot too in many cases, which is luminous and contributes to gas radiation.

Origin of studies on gas radiation

The origin of all this is from the classic problem of radiation from molten glass in a furnace. Glass blowing is a very traditional and old technique. When temperature distribution in molten glass was measured, it was found to be more uniform than that expected from heat conduction alone. When researchers wrote the equations and solved for the temperature distribution using heat conduction alone, they found more variation than was actually measured and they started investigating why it was so. They thought convection may be the culprit and tried incorporating it. But this too did not help the simulations better agree with the measured results. It was later discovered that glass itself has emission and absorption characteristics and one has to integrate conduction with the equation of radiation. In the late 1940s, it became clear that gas radiation was largely responsible for this behaviour, and it was observed that when radiation interacts with a substance, part of the energy may be redirected by scattering, which may in turn, be caused by a small particle such as an electron or a huge one such as a planet. So a wide range of length scales is involved.

Depending on the length scale, different theories of scattering are used. If the length scale is very small, Rayleigh scattering is adequate. If we consider ice, water or rain in the atmosphere, we cannot use the Rayleigh scattering, but instead use the Lorentz–Mie scattering theory. However, if the particles are big enough and are not spherical, this theory will also not work and geometric optics has to be used.

5.1 Principal Difficulties in Studying Gas Radiation

In gas radiation, everything is happening from a volume and not from a surface like before. Absorption, emission and scattering are a function of the wavelength, λ and happen at all locations within the medium, which makes it mathematically very difficult. Spectral effects are more pronounced in gases than from solid surfaces. This gray gas assumption is more a myth than a reality. Hence, the engineering treatment of gas radiation would involve simplification of one or both of these difficulties.

5.2 Important Properties for Study of Gas Radiation

- κ_λ is a monochromatic or spectral absorption coefficient (m^{-1}) Therefore, if incident monochromatic radiation is given by I_λ, the absorption by the gas per unit volume per unit solid angle per unit wavelength interval is given by $\kappa_\lambda I_\lambda$ in $W/m^3.\mu m.sr.$
- Similarly, we have ϵ_λ which is the monochromatic or spectral emission coefficient which also has the unit m^{-1}. The ϵ_λ we define for gas radiation is different from the ϵ_λ we defined for radiation from surfaces. The emission by the gas is given by $\epsilon_\lambda I_\lambda(T_g)$ $W/m^3.\mu m.sr.$

5.3 Equation of Transfer or Radiative Transfer Equation (RTE)

Consider a gas volume with a cross-sectional area dA and thickness ds. $I_{\lambda,s}$ is the incoming radiation in direction s, while $I_{\lambda,s+ds}$ is the outgoing radiation as shown in Fig. 5.3. The area dA is normal to the direction s such that the radiation is travelling in a direction normal to the cross-sectional area. We are now trying to find out the rate of change of the intensity of the radiation as it passes through the gas volume. Then, we need to determine the factors which make this rate of change of intensity not equal to zero. Since we are neglecting scattering, there can be only two phenomena, namely absorption within the gas volume or re-emission from the gas volume. The balance between this emission and absorption will lead to this dI_λ. It looks very simple and unassuming, but this is only deceptively innocuous.

The change in intensity $I_{\lambda,s}$ when passing through the gas volume

$$= I_{\lambda,s+ds}.dA - I_{\lambda,s}.dA \qquad (5.1)$$

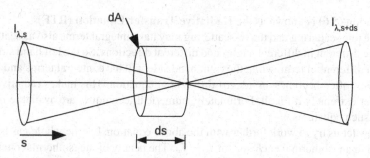

Fig. 5.3 Gas volume used in the derivation of the RTE

By Taylor's series expansion,

$$I_{\lambda,s+ds}dA - I_{\lambda,s}dA = \frac{dI_\lambda}{ds}ds.dA + \frac{d^2I_\lambda}{ds^2}\frac{ds^2}{2!}.dA \qquad (5.2)$$

We set the higher order terms to 0, which is the error associated with this approximation.
Energy absorbed by the gas in the interval $d\lambda$ is

$$= \kappa_\lambda I_\lambda dAds \qquad (5.3)$$

Energy emitted by the gas volume:

$$\epsilon_\lambda I_{b\lambda}(T_g)dAds \qquad (5.4)$$

The intensity I_λ, which is incoming, need not be related to the temperature T_g, as it can come from anywhere. But what it emits will be completely dependent on its temperature T_g.

With the absorption, the I_λ will tend to decrease as the radiation passes through the gas volume but this will be compensated by the emission from the gas. So now we have to do the energy balance for the gas volume,

$$\frac{dI_\lambda}{ds} = \epsilon_\lambda I_{b\lambda}(T_g) - \kappa_\lambda I_\lambda \qquad (5.5)$$

We can cancel $dAds$ throughout, implicitly assuming that $dAds \neq 0$. Hence, these equations are not valid at mathematical points, which have no area or volume. These are valid only around a small area or volume.

On doing this we get the following

$$\frac{dI_\lambda}{ds} + \kappa_\lambda I_\lambda = \epsilon_\lambda I_{b\lambda}(T_g) \qquad (5.6)$$

Equation (5.6) is known as the **Radiative Transfer Equation (RTE)**.

The out-scattering and the in-scattering may have integral terms also as scattering may be different in different angles and different directions. So the left hand side will have a differential term, while the right hand side will have integral terms and hence Eq. (5.6) often becomes an integro-differential equation. This makes radiative heat transfer extremely difficult. Fortunately, numerical techniques are available now to solve such equations.

Now let us try to work further with the above equation. Let the whole gas be contained in an isothermal enclosure at $T = T_g$. The beauty of the isothermal enclosure is that the gas and the walls are all at the same temperature T_g. The radiation coming out from this enclosure will be analogous to radiation coming from a black body. Therefore, if we take a sample and find out the I_λ in a particular wavelength interval and if we take another sample, both will be same. Therefore, $\frac{dI_\lambda}{ds} = 0$ everywhere within the isothermal cavity.

$$I_\lambda \neq f(s) \tag{5.7}$$
$$I_\lambda = I_{b,\lambda}(T_g) \tag{5.8}$$

Substituting in Eq. 5.6

$$0 + \kappa_\lambda I_{b\lambda}(T_g) = \epsilon_\lambda I_{b\lambda}(T_g) \tag{5.9}$$
$$\therefore \kappa_\lambda = \epsilon_\lambda \tag{5.10}$$

This is Kirchhoff's law.

The isothermal enclosure concept is used only to prove this but it is universally applicable; we can make measurements and check it. Getting back to the RTE, we have

$$\frac{dI_\lambda}{ds} + \kappa_\lambda I_\lambda = \kappa_\lambda I_{b\lambda}(T_g) \tag{5.11}$$

Equation 5.11 looks so simple, but when we try to solve the equation, we will see that it will lead to an integral that cannot be solved.

Let us now consider an asymptotic case where the absorption is more important than the emission. Consider a wall which is black at a temperature T_w, surrounded by gas at T_g. Let a receiver be placed at a distance L from the wall which receives the radiation coming out of the wall. This radiation is made up of two contributions

1. The radiation from the wall which is modified or attenuated by the participating medium, which eventually reaches the receiver.
2. Radiation from the gas also falls on the receiver. The radiation on the receiver can come from any portion of the wall, and hence, we have to derive the formulation for a general angle θ ($\theta = 0$, is the special case).

When $T_w \gg T_g$, $\epsilon_\lambda I_{b\lambda}(T_g)$ is much smaller compared to the other terms in the equation because the emission component is very less. If we have a 1000 K wall while the gas is only at 300 K then this condition is approached. Then the RTE becomes (if x is taken as the s direction),

$$\frac{dI_\lambda}{dx} + \kappa_\lambda I_\lambda = 0 \tag{5.12}$$

$$\frac{dI_\lambda}{I_\lambda} = -\kappa_\lambda dx \tag{5.13}$$

Assuming a gray gas,

$$\frac{dI}{I} = -\kappa dx \tag{5.14}$$

If we integrate this, we get $I = Ae^{-kx}$.

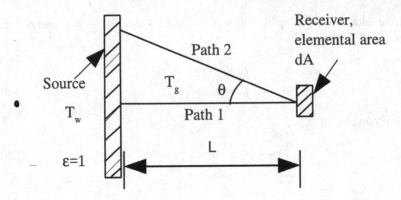

Fig. 5.4 Radiation heat flux at a distance L from the source in a participating gas

At $x = 0$, $I = I_0$, $\therefore A = I_0$, Now the solution becomes $I = I_0 e^{-kx}$.

Therefore, the radiation decays exponentially as it passes through the depth. This is called **Beer's law or Lambert's law or Beer–Lambert's law**.

Now consider solar radiation which is impinging on the ocean waters. Let us say it is I_0 at the surface and as it goes in, it will go as $I_0 e^{-kx}$. As we go to the bottom layers of the ocean, the radiation received by the bottom layers is much less compared to the top. Because of this, the bottom layers are at a temperature much lower than that of the top layers. If the top layer is at 30 °C, then at 1km depth, it may be at 10 °C. This is a stable temperature gradient because the warmer and lighter water stays at the top. This is what helps maintain the aquatic life. We can see that it is very difficult for organisms to survive very deep below because light is not available for photosynthesis. Now as engineers, if we want to exploit this temperature difference to run a heat engine, we are talking about what is called the Ocean Thermal Energy Conversion or OTEC.

So, simple radiative transfer equations can be used to explain so many phenomena! In the case of above example the source is at 5800 K, and we do not worry about $\epsilon_\lambda I_{b\lambda}(T_g)$ as the medium is around 300 K or so.

We will look at the simplified treatment of the RTE equation and some of its solutions. Consider a black wall whose emissivity $\epsilon = 1$ at temperature T_w. Next to it is a plane gas layer of thickness L, which is infinite in the other two directions at temperature T_g. Essentially, we are looking at how the intensity varies in the direction x. Now, we want to solve the equation of transfer and find out how I propagates with x.

$$\frac{dI_\lambda}{ds} + \kappa_\lambda I_\lambda = \kappa_\lambda I_{b\lambda}(T_g) \tag{5.15}$$

If s is oriented at an angle θ to dA (look at path 2 in Fig. 5.4)

$$\cos(\theta)\frac{dI_\lambda}{ds} + \kappa_\lambda I_\lambda = \kappa_\lambda I_{b\lambda}(T_g) \tag{5.16}$$

When $\theta = 0$, equation reduces to Eq. (5.15). The radiation arriving at the area dA consists of two components.

1. The first is the radiation from the wall, which goes through the gas gets attenuated (because the gas is absorbing radiation and participating in the process) and arrives at dA.
2. The second component arriving at dA is because of emission from the gas.

So the two components are transmission by the gas and emission from the gas. For a plane gas layer,

$$\frac{dI^+}{dx} + \kappa I^+ = \frac{\kappa \sigma T_g^4}{\pi} \tag{5.17}$$

We have done several things here. We are assuming that the gas is gray as we knocked out κ_λ and replaced it by κ. We have used I^+, because we are looking at the positive direction of x. We have replaced $I_{b,\lambda}$ by I and ds by dx. We will solve the equation of transfer for the straight path and infer what the solution for the slant path will be.

5.4 Solution for the Straight Path

There will be two parts to the solution, the complementary function (CF) and the particular integral (PI). The complementary function is obtained by setting the right hand side to 0 of Eq. 5.17.

$$\text{General solution} = \text{CF} + \text{PI}$$

To get the CF, we do the following

$$\frac{dI^+}{dx} + \kappa I^+ = 0 \tag{5.18}$$

$$I^+ = Ae^{-\kappa x} \tag{5.19}$$

The PI, $I^+ = \frac{\sigma T_g^4}{\pi}$ is evident. The general solution is then

$$I^+ = Ae^{-\kappa x} + \frac{\sigma T_g^4}{\pi} \tag{5.20}$$

At $x = 0$, $I^+ = \frac{\sigma T_g^4}{\pi}$. Therefore,

$$\frac{\sigma T_w^4}{\pi} = Ae^0 + \frac{\sigma T_w^4}{\pi} \tag{5.21}$$

$$A = \frac{\sigma}{\pi}(T_w^4 - T_g^4) \tag{5.22}$$

$$I^+(x) = \frac{\sigma T_w^4}{\pi}e^{-\kappa x} + \frac{\sigma T_g^4}{\pi}(1 - e^{-\kappa x}) \tag{5.23}$$

At $x = L$

$$I^+(L) = \frac{\sigma T_w^4}{\pi}e^{-\kappa L} + \frac{\sigma T_g^4}{\pi}(1 - e^{-\kappa L}) \tag{5.24}$$

The units of κL is m^{-1}.m, a dimensionless quantity. This frequently appears in radiation heat transfer and is called the **optical depth** (τ). If the optical depth of one medium is higher than that of another medium, it means that its capacity to absorb gas radiation is much more. An optically thin gas is one in which τ is very small, and it absorbs a small amount of the incident radiation. An optically thick gas is something that has a high τ and little will emerge out of it. τ is a dimensionless quantity. In general,

$$I^+(x) = \frac{\sigma T_w^4}{\pi}e^{-\tau_x} + \frac{\sigma T_g^4}{\pi}(1 - e^{-\tau_x}) \tag{5.25}$$

The performance metric for the gas layer of thickness L is the ratio of the intensity at x to the intensity at 0, given by

$$\frac{I^+(x)}{I^+(0)} = e^{-\tau_x} + \left(\frac{T_g}{T_w}\right)^4 (1 - e^{-\tau_x}) \tag{5.26}$$

If $\frac{T_g}{T_w} \ll 1$, (for example, the gas temperature is 300 K and the wall temperature is 1500 K), the above equation reduces to Beer–Lambert's law.

$$\frac{I^+(x)}{I^+(0)} = e^{-\tau_x} \tag{5.27}$$

A good question to ask now is Why should we always worry about the straight path as radiation can arrive at the elemental area dA from anywhere? We did not want to work with the general path and get into a mess. Now having got the result for simple path, by induction or inference, we can get what I^+ will be for a slant path. For a slant path,

$$I^+\left(\frac{\tau_x}{\cos\theta}\right) = \frac{\sigma T_w^4}{\pi}e^{-\frac{\tau_x}{\cos\theta}} + \frac{\sigma T_g^4}{\pi}(1 - e^{-\frac{\tau_x}{\cos\theta}}) \tag{5.28}$$

Let $\cos\theta = \mu$

$$I^+\left(\frac{\tau_x}{\mu}\right) = \frac{\sigma T_w^4}{\pi}e^{-\frac{\tau_x}{\mu}} + \frac{\sigma T_g^4}{\pi}(1 - e^{-\frac{\tau_x}{\mu}}) \tag{5.29}$$

We can check for the asymptotic correctness of the expression by making $\cos\theta = \mu = 1$.

This is the solution for the equation of transfer. We can see that even for the simple case of one wall being black, a gas layer which is gray, and for one positive direction of x, it looks quite formidable. Even so, this can be handled by pen and paper; the more complex ones have to be solved using programs.

5.5 Heat Fluxes

The heat flux at x = 0, going out in the positive direction of x is given by

$$q_L^+ = \int_{\phi=0}^{2\pi}\int_{\theta=0}^{\frac{\pi}{2}} I_L^+(\mu)\cos\theta\sin\theta d\theta d\phi \tag{5.30}$$

I_L is only a function of μ and cannot be pulled out of the integral, and it is not a function of x as we are specifically evaluating the integral at $x = L$. If we have azimuthal symmetry, then integration with respect to $d\phi$ can be done. The first simplification is to pull the $d\phi$ out of the integral, as follows:

$$q_L^+ = 2\pi\int_0^1 I_L^+(\mu)\cos\theta d(\cos\theta) \tag{5.31}$$

Here, the $\sin\theta$ was taken as $-d(\cos\theta)$ and, therefore, the limits were changed from 1 to 0 into 0 to 1. Now, $d(\cos\theta)$ can be written as $d\mu$, and we have to substitute for $I_L(\mu)$ and accomplish the integration.

$$q_L^+ = 2\pi\int_0^1 I_L^+(\mu)\mu d(\mu) \tag{5.32}$$

$$q_L^+ = 2\pi\int_0^1 \left\{\frac{\sigma T_W^4}{\pi}\exp\left(-\frac{\tau_L}{\mu}\right) + \frac{\sigma T_g^4}{\pi}\left(1 - \exp\left(-\frac{\tau_L}{\mu}\right)\right)\right\} \tag{5.33}$$

$$q_L^+ = 2\pi \left[\frac{\sigma T_g^4}{\pi} \frac{\mu^2}{2} \right]_0^1 + \int_0^1 2\pi \frac{\sigma}{\pi} (T_W^4 - T_g^4) \exp\left(-\frac{\tau_L}{\mu} \right) \mu d\mu \qquad (5.34)$$

$$q_L^+ = \sigma T_g^4 + 2\sigma (T_W^4 - T_g^4) \int_0^1 \exp\left(-\frac{\tau_L}{\mu} \right) \mu d\mu \qquad (5.35)$$

Whatever method we try, this integration cannot be done. This is an integral that frequently appears in radiative heat transfer. It is called an **exponential integral of order 3**.

5.5.1 Exponential Integral of Order n

The expression for an exponential integral of order "n" is given by

$$E_n(t) = \int_0^1 \mu^{n-2} \exp\left(-\frac{t}{\mu} \right) d\mu \qquad (5.36)$$

When n=3, we have

$$E_3(t) = \int_0^1 \mu^1 \exp\left(-\frac{t}{\mu} \right) d\mu \qquad (5.37)$$

μ is basically a dummy variable.

5.5.2 Salient Properties of $E_3(x)$

$$\lim_{t \to 0} E_3(t) = \left(\frac{1}{2} - t \right) \qquad (5.38)$$

$E_3(0) = 0.5$. The t in Eq. (5.37) corresponds to actually τ or optical depth. If the optical depth is very small, this exponential integral reduces to $\frac{1}{2} - t$. So, the limit where t approaches 0 is called the **optically thin limit for radiation**. It is optically thin enough to allow the approximation, but it is not optically thin enough to neglect gas radiation. $E_3(\infty) = 0$. Now, we can write the general expression for q_L^+ as

Table 5.1 Values of exponential integral $E_3(x)$

x	$E_3(x)$	x	$E_3(x)$
0.00	0.50000	0.60	0.19156
0.01	0.49029	0.65	0.17830
0.02	0.48098	0.70	0.16607
0.03	0.47201	0.75	0.15477
0.04	0.46333	0.80	0.14433
0.05	0.45493	0.85	0.13466
0.06	0.44677	0.90	0.12571
0.07	0.43884	0.95	0.11741
0.08	0.43113	1.00	0.10970
0.09	0.42362	1.20	0.08394
0.10	0.41630	1.40	0.06458
0.15	0.38228	1.60	0.04991
0.20	0.35195	1.80	0.03872
0.25	0.32469	2.00	0.03014
0.30	0.30005	2.25	0.02212
0.35	0.27768	2.50	0.01630
0.40	0.25729	2.75	0.01205
0.45	0.23867	3.00	0.00893
0.50	0.22161	3.25	0.00664
0.55	0.20595	3.50	0.00495

$$q_L^+ = \sigma T_g^4 + 2\sigma(T_W^4 - T_g^4)E_3(\tau_L) \tag{5.39}$$
$$= 2E_3(\tau_L)\sigma T_W^4 + \sigma T_g^4[1 - 2E_3(\tau_L)] \tag{5.40}$$

The values of $E_3(x)$ for various values of x are presented in Table 5.1.

For the optically thin gas,

$$E_3(\tau_L) = \frac{1}{2} - \tau_L \tag{5.41}$$
$$\text{and } 1 - 2E_3(\tau_L) = \tau_L \tag{5.42}$$
$$\therefore \quad q_L^+ = 2\tau_L\sigma T_g^4 + (1 - 2\tau_L)\sigma T_w^4 \tag{5.43}$$

$\therefore$ The radiation arriving at $x = L$ consists of two parts namely

1. the radiation which is directly coming from the gas (first term)
2. the radiation which is coming from the wall and is attenuated by the gas (second term)

$$\epsilon_g = 2\tau_L = (2L)(\kappa) \tag{5.44}$$

$$\tau_g = 1 - 2\tau_L = (1 - \epsilon_g) \tag{5.45}$$

$$\therefore q_L^+ = \epsilon_g \sigma T_g^4 + \tau_g \sigma T_w^4 \tag{5.46}$$

In the above two equations, ϵ_g and τ_g can be considered to be the emissivity of the gas and gas transmissivity respectively.

At the end, we are able to define the gas emissivity which consists of two parts $2L$ and κ. $2L$ is basically related to the geometry while κ is related to the capacity of the gas to absorb. Therefore, when we combine the geometry part and the thermal part, we are able to get the equivalent gas emissivity, which we can use with the radiosity formulation developed for the evacuated enclosure, by modifying it.

This $2L$ represents the mean path travelled by all rays to arrive at the elemental area dA which is located exactly at a minimum distance L from the wall. For $\cos\theta = 1$, it will be just L. For all the others, it will be $L/\cos\theta$ and hence keeps changing. This $2L$ is some sort of an average or mean length which a ray travels before hitting the elemental area in the receiver (Fig. 5.4). Hence, $2L$ is called the **mean beam length**, usually denoted by L_e.

We started out with the equation of transfer and now the formulation has reached a critical stage, where the gas emissivity is a product of two distinct parts, wherein the thermal part can be completely separated from the geometry part. This $2L$ is the mean beam length for a plane gas layer. This mean beam length will change for a cylinder, sphere and so on. So, if we are able to calculate the mean beam length and we also know κ, we can calculate the gas emissivity and from that, the gas absorptivity can be determined and we can proceed further.

Let us consider a hemispherical gas volume whose radius is R (see Fig. 5.5). There is an elemental area dA at the centre of the bottom surface. We are looking at this gas volume which is absorbing and emitting. We also have a small area on the hemisphere, and we are "connecting" to the elemental area dA by trying to find out what the radiation arriving from here at dA after travelling through the gas volume is.

The gas is at a temperature T_g and is optically thin. These are the two assumptions. The idea behind this exercise is to understand the physics behind the mean beam length concept. We have derived that for an optically thin gas, for radiation from somewhere arriving from length L, expression for intensity for $I^+(L)$, (where $+$ indicates the positive direction of x) is

Fig. 5.5 Hemispherical gas volume used for elucidating the concept of mean beam length

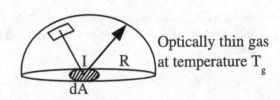

Optically thin gas at temperature T_g

$$I^+(L) = \frac{\sigma T_W^4}{\pi} \exp(-\kappa L) + \frac{\sigma T_g^4}{\pi}(1 - \exp(-\kappa L)) \tag{5.47}$$

Now let us consider q^- instead of I^+,

$$q^-(R) \approx \pi \left[\frac{\sigma T_W^4}{\pi} \exp(-\kappa R) + \frac{\sigma T_g^4}{\pi}(1 - \exp(-\kappa R)) \right] \tag{5.48}$$

We have done two things here:

1. instead of L, we have used R
2. we have multiplied the intensity expression by π because it is isotropic.

It is the gas which is contributing more to the radiation. T_w may be at 300 K while the gas may be at 1500 K. For such a case, we can neglect the first term and so Eq. (5.48) becomes

$$q^-(R) = \sigma T_g^4(1 - \exp(-\kappa R)) \tag{5.49}$$

For optically thin gases, $\kappa R << 1$, so the expression within brackets can be expanded as κR.

$$q^-(R) = \sigma T_g^4 \kappa R \tag{5.50}$$

If we are able to write the flux as σT^4 multiplied by some quantity (κR in this case), this quantity in brackets can be called as **equivalent emissivity**.

$$q^-(R) = \epsilon_g \sigma T_g^4 \tag{5.51}$$
$$\text{where, } \epsilon_g = \kappa R \tag{5.52}$$

In Eq.5.52, κ is thermal part and R is the geometric part. When we considered a plane wall which was black and at a temperature T_w surrounded by an isothermal gas at T_g, we figured out that the mean beam length was equal to $2L$. Here, the mean beam length is R itself.

However, these analytical ways of deriving the mean beam length cannot be done for each and every geometry. Certain mean beam lengths have to be calculated or evaluated. Tabulated values of the mean beam length for a few geometries are given in Table 5.2, which may be used in problems involving those geometries.

In this table, the mean beam length between two parallel plates is $1.8L$ instead of 2L because of the expansion of $E_3(T)$ in the vicinity of T. For an arbitrary shape of volume V and surface area A, the mean beam length $= 3.6\frac{V}{A}$. If we apply this formula for the plane gas layer of thickness L, we get $1.8L$.

Table 5.2 Mean beam lengths L_e for different gas geometries (adapted from Howell et al. (2011)

Sl. No.	Geometry	Characteristic length	L_e
1.	Hemisphere radiating to element at centre of base	Radius R	R
2.	Sphere (radiation to surface)	Diameter D	0.65D
3.	Infinite circular cylinder (radiation to curved surface)	Diameter D	0.95D
4.	Semi-infinite circular cylinder (radiation to base)	Diameter D	0.65D
5.	Circular cylinder of equal height and diameter (radiation to entire surface)	Diameter D	0.60D
6.	Infinite parallel planes (radiation to planes)	Spacing between planes L	1.80L
7.	Cube (radiation to any surface)	Side L	0.66L
8.	Arbitrary shape of volume V (radiation to surface of area A)	Volume to area ratio V/A	3.6V/A

Example 5.1 Consider a gray gas with an absorption coefficient of $\kappa = 0.15$ m^{-1}. It is maintained at a temperature of 400 K and is 0.4 m thick (Fig. 5.6). A black wall at 500 K is at $x = 0$. Determine the intensity at $x = 0.4$ m for

- straight path
- slant path at 60°

Use the small τ approximation as well as the exponential integral for evaluating the heat flux at $x = 0.4$ m and comment on the result.

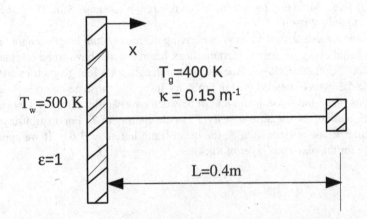

$T_w = 500$ K

$\varepsilon = 1$

X

$T_g = 400$ K

$\kappa = 0.15$ m^{-1}

L=0.4m

Fig. 5.6 Problem geometry for Example 5.1

Solution:

(a)

$$I_L^+(\theta = 0) = \frac{\sigma T_W^4}{\pi} \exp(-\kappa L) + \frac{\sigma T_g^4}{\pi}(1 - \exp(-\kappa L)) \quad (5.53)$$

$$I_{L=0.4}^+(\theta = 0) = 1062.3 + 26.91 \quad (5.54)$$

$$I_{L=0.4}^+(\theta = 0) = 1089.21 \text{ Wm}^{-2}sr \quad (5.55)$$

The intensity coming out of the wall is

$$I_0^+ = \frac{\sigma T_W^4}{\pi} = 1128.01 \text{ Wm}^{-2}sr \quad (5.56)$$

The intensity coming out of the wall is not asked in the problem but we have calculated this to probe deeper into the result. Now look at the ratio of I at $x = L$ and I at $x = 0$. It is almost 0.98 or 0.99. This shows that the gas is not absorbing much. Therefore, it is an optically thin gas. If this ratio were instead got as 0.5 or 0.6, then the gas would no longer be thin.

$$I_L^+(\theta = 60) = \frac{\sigma T_W^4}{\pi} \exp\left(-\frac{\kappa L}{\cos(60)}\right)$$
$$+ \frac{\sigma T_g^4}{\pi}\left(1 - \exp\left(-\frac{\kappa L}{\cos(60)}\right)\right) \quad (5.57)$$

$$I_{L=0.4}^+(\theta = 60) = 1052.7 \text{ Wm}^{-2}sr \quad (5.58)$$

(b) Small τ approximation:

$$\tau_L = \kappa L \quad (5.59)$$
$$= 0.15 \times 0.4 = 0.06 \quad (5.60)$$
$$E_3(\tau_L) = 0.44 \quad (5.61)$$
$$q_L = 5.67 \times 10^{-8}[400^4(1 - 0.88) \quad (5.62)$$
$$+ 500^4 \times 0.88 \quad (5.63)$$
$$= 3292.7 \text{ Wm}^{-2} \quad (5.64)$$
$$q_L(EI) = \sigma T_g^4[1 - 2E_3(\tau L)] + 2\sigma T_W^4(E_3(\tau L)) \quad (5.65)$$

From the tables,

$$E_3(0.06) = 0.447 \quad (5.66)$$

$$q_L(EI) = 5.67 \times 10^{-8}(400^4)(1 - 2 \times 0.447) \qquad (5.67)$$
$$+ 2 \times 5.67 \times 10^{-8}(500^4) \times 0.447 \qquad (5.68)$$
$$= 3322 \text{ Wm}^{-2} \qquad (5.69)$$
$$\text{Percentage error} = \frac{(3321 - 3293)}{3321} \times 100 = 0.87 \qquad (5.70)$$

If $\tau < 0.1$, then the optical thin gas approximation is good. So whether a gas is optically thin or thick depends on not only its absorptivity but also on the length scale involved. It is the product which matters. One can have a very heavily absorbing gas but if the length scale is only 1 mm or 2 mm, the medium will still be optically thin. On the contrary, one can encounter a poorly absorbing gas but because of a large thickness, κL product can be significant.

In the previous problem, we have made many assumptions—a gray gas, single gas, isothermal gas. Any of these assumptions can be questioned. Most importantly, this analysis cannot be used in an internal combustion engine or a combustion chamber simply because there is never a single gas there. So, the next levels of complexity will be

(1) How do we get the ϵ_g for a mixture of gases?
(2) What will happen if the walls are not black?

Are there some recipes available if there is a mixture of gases? Yes! The commonly encountered participating gases are carbon dioxide and water vapour, as found in power plants, internal combustion engines. Several researchers have done experiments at 1 atmospheric pressure and have given charts for a mixture of carbon dioxide and water vapour, using which we can calculate the mixture properties.

Example 5.2 Two infinitely long vertical plates are parallel to each other (Fig. 5.7). Both the plates are black and are at temperatures $T_1 = 1500$ K and $T_2 = 900$ K, respectively. The spacing between the two plates is 1.5 m and is filled with a gray gas at 1200 K with an absorption coefficient of 0.08 m^{-1}. Determine the heat transfer rate at each of the two boundaries.

Solution:

We can use superposition solving,for the left and right walls separately. For one wall, from left to right we consider in the positive x direction and for the other we consider the negative x direction. We need to take the algebraic sum of the heat fluxes. The heat flux expression must be asymptotically correct, such that when $\kappa = 0$, we must get the original parallel plate formula for

$$q = \sigma(T_1^4 - T_2^4) \text{ when } \epsilon_1 = \epsilon_2 = 1$$

Fig. 5.7 Problem geometry
for Example 5.2

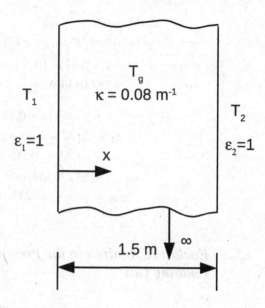

Left wall:

Heat flux is transferred from left wall to the right wall.

$$q_0 = 2\sigma T_1^4 E_3(\tau_x) + \sigma T_g^4 [1 - 2E_3(\tau_x)] \qquad (5.71)$$

Right wall:

$$q_L = 2\sigma T_2^4 E_3(\tau_L - \tau_x) + \sigma T_g^4 [1 - 2E_3(\tau_L - \tau_x)] \qquad (5.72)$$

Heat flux is transferred from the right wall to the left wall.

$$\text{Net radiative flux at x} = q_0 + q_L \qquad (5.73)$$
$$\text{Special case: } \kappa = 0 \qquad (5.74)$$
$$q_0 = \sigma T_1^4 \qquad (5.75)$$
$$q_L = \sigma T_2^4 \qquad (5.76)$$
$$\text{Net radiative flux at x} = q_0 - q_L = \sigma(T_1^4 - T_2^4) \qquad (5.77)$$

The contribution for the gas volume term is 0 from both the left and the right. This is consistent with our understanding of the problem. Now, substituting the numerical values in the equations for the problem under consideration, we get q_0 and q_L. At $x = 0$,

$$E_3(0) = 0.5 \tag{5.78}$$

$$\begin{aligned} q_0 &= 2\sigma T_1^4 E_3(\tau_x) + \sigma T_g^4[1 - 2E_3(\tau_x)] \\ &= 2 \times 5.67 \times 10^{-8} \times 1500^4 \times 0.5 + 0 \\ &= 287.043 \text{ kW/m}^2 \end{aligned} \tag{5.79}$$

$$E_3(\tau_L - \tau_x) = E_3(0.12) = 0.40 \tag{5.80}$$

$$\begin{aligned} q_L &= 2\sigma T_2^4 \times 0.40 + \sigma T_g^4[1 - 2 \times 0.40] \\ &= 29761 + 23515 \\ &= 53.276 \text{ kW/m}^2 \end{aligned} \tag{5.81}$$

$$q_{net} = q_0 - q_L = 233.767 \text{ kW/m}^2 \tag{5.82}$$

5.5.3 Enclosure Analysis in the Presence of an Absorbing or Emitting Gas

Let us consider two areas A_1 and A_2 in an enclosure and two elemental areas in them, dA_1 and dA_2 (Fig. 5.8). The distance between them is R. The unit vectors are n_1 and n_2 and the angles subtended by them are θ_1 and θ_2, respectively. What is the difference between this R and one we considered in the previous chapter? This R is the path through the gas which interferes with the radiation passing through, while previously R was the path through vacuum or a non participating medium.

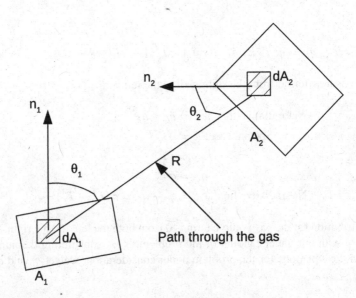

Fig. 5.8 Enclosure analysis in the presence of an absorbing and emitting gas

All the surfaces in the enclosure are gray and diffuse. The gas is optically thin. So, the radiation leaving dA_1 that falls on dA_2 is given by

$$\frac{J_1 dA_1 dA_2 \cos(\theta_1) \cos(\theta_2)}{\pi R^2} \exp(-\kappa R) \tag{5.83}$$

The extra term in expression (5.83) is the attenuation which is exponential. For an optically thin gas,

$$\exp(-\kappa R) \approx 1 - \kappa R \tag{5.84}$$

Substituting Eq. (5.84) in Eq. (5.83), we have

$$\frac{J_1 dA_1 dA_2 \cos(\theta_1) \cos(\theta_2)}{\pi R^2} - \kappa \frac{J_1 dA_1 dA_2 \cos(\theta_1) \cos(\theta_2)}{\pi R} \tag{5.85}$$

Therefore, the irradiation of A_2 because of the radiation emanating from A_1 is given by the double integral of the two terms over A_1 and A_2. Assuming uniform radiosity, we take the J_1 out.

The irradiation of A_2 because of radiation from A_1 is given by

$$J_1 \kappa \int_{A_1} \int_{A_2} \frac{dA_1 dA_2 \cos(\theta_1) \cos(\theta_2)}{\pi R^2} - J_1 \kappa \int_{A_1} \int_{A_2} \frac{dA_1 dA_2 \cos(\theta_1) \cos(\theta_2)}{\pi R} \tag{5.86}$$

The term within the first integral is $A_1 F_{12} = A_2 F_{21}$. The term within the second integral should also be something similar and let us say it can be given by $A_1 L_{12} = A_2 L_{21}$. The irradiation on A_2 only from the radiation coming from A_1 is given by

$$J_1 A_1 F_{12} - \kappa J_1 A_1 L_{12} = A_2 G_2^+ \tag{5.87}$$

Using reciprocal rules, we can write

$$A_2 G_2^+ = J_1 A_2 F_{12} - \kappa J_1 A_2 L_{21} \tag{5.88}$$

$$G_2^+ = J_1 F_{21} \left[1 - \frac{\kappa L_{21}}{F_{21}} \right] = J_1 F_{21} \left[1 - \frac{\kappa L_{12}}{F_{12}} \right] \tag{5.89}$$

By the same token,

$$G_1^+ = J_2 F_{12} \left[1 - \frac{\kappa L_{21}}{F_{21}} \right] \tag{5.90}$$

For a two surface enclosure problem, please remember the parallel plate formula for an evacuated case, the G_1 will be J_2 multiplied by the view factor. Now, in this case, we get J_2 multiplied by the view factor and another factor. This extra factor can be deemed to be the transmittance of the gas. If the transmittance of the gas is 1, the expression reduces to what we obtained for the evacuated enclosure or the transparent medium. So whatever is within the brackets, will have a value less than 1 consequent upon the fact that $\kappa \neq 0$. κ represents the amount of absorption. Even if the gas is optically thin, κ will still have a value. Therefore, $1 - \frac{\kappa L_{21}}{F_{21}}$ can be considered as $\tau(g)$ or the transmissivity of the gas.

$$G_1^+ = J_2 F_{12} \tau_{12} \tag{5.91}$$

$$L_{12} = L_{21} = \text{mean beam length} \tag{5.92}$$

L_{ij} also follows the reciprocal rule.

$$\tau_{12} = 1 - \frac{\kappa L_{21}}{F_{21}} = 1 - \frac{\kappa L_{12}}{F_{12}}. \tag{5.93}$$

Having defined this, we have to modify the irradiation terms. G_i consists of two components

1. Irradiation from other surfaces, which can be handled by calculating τ_{ij} for all the surfaces
2. Gas emission, given by $\sigma \epsilon_g T_g^4$

$$G_i = \epsilon_g \sigma T_g^4 + \sum_{j=1}^{N} F_{ij} J_j \tau_{ij} \tag{5.94}$$

$$\text{where } \tau_{ij} = \left(1 - \frac{\kappa L_{ij}}{F_{ij}}\right) = \left(1 - \frac{\kappa L_{ji}}{F_{ji}}\right) \tag{5.95}$$

There are two critical changes, we have done to irradiation with respect to evacuated enclosures:

• We have an emission term in the irradiation too.
• Within the summation, we have the gas transmittance included by the use of τ_{ij}.

The gas transmittance consists of L_{12}, which is the simplified representation of the solution to the equation of transfer. If it is not an optically thin gas, all these tricks do not work, and we have to solve the equation of transfer and do full blown radiation calculations. Just like for the case of a lumped capacitance system in conduction analysis, wherein we say the whole body is at one temperature, here too, the approximations are valid only for a single optically thin gas.

$$J_i = \epsilon_i \sigma T_i^4 + (1 - \epsilon_i) G_i \tag{5.96}$$

$$q_i = J_i - G_i \tag{5.97}$$

So the formulation is exactly as before, except that the irradiation has extra terms and one modified term. The extra term is the one involving emission while the modified term is $F_{ij} J_j$.

This is basically the theory of evacuated enclosure applied to an absorbing/emitting gas. If $\kappa = 0$ and the gas is not participating, $\tau_{12} = 1$, $\epsilon_g = 0$ and hence the first term

Fig. 5.9 Geometry and pertinent data for example 5.3

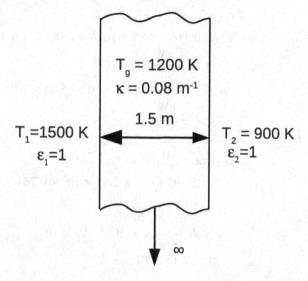

$T_g = 1200$ K
$\kappa = 0.08$ m^{-1}

1.5 m

$T_1 = 1500$ K
$\varepsilon_1 = 1$

$T_2 = 900$ K
$\varepsilon_2 = 1$

∞

in Eq. (5.94) becomes 0 while G_i becomes essentially the summation of just $F_{ij} J_j$. Hence, there is an asymptotic correctness associated with this.

Example 5.3 Revisit Example 5.2, and solve it using the modified enclosure theory (Fig. 5.9).

Solution:

View factors:
$F_{11} = 0$, $F_{12} = 1$, $F_{21} = 1$, $F_{22} = 0$

$$\tau_{12} = \tau_{21} = 1 - \frac{\kappa L_{12}}{F_{12}} \tag{5.98}$$

$$\tau_{12} = 1 - \frac{(0.08 \times 1.8 \times 1.5)}{1} = 0.784 \tag{5.99}$$

What does this 0.784 represent physically? It means that the gas is allowing 80% of the radiation to go through. It is absorbing 20% of the radiation.

$$\epsilon_g = 1 - \tau_{12} = 0.216 \tag{5.100}$$

$$J_1 = \epsilon_1 \sigma T_1^4 + (1 - \epsilon_1) F_{12} J_2 = \sigma T_1^4$$

$$J_1 = 2.87 \times 10^5 \ \frac{W}{m^2} \tag{5.101}$$

$$J_2 = \epsilon_2 \sigma T_2^4 + (1 - \epsilon_2) F_{21} J_1$$

$$= 37200 \ \frac{W}{m^2} \tag{5.102}$$

$$G_1 = \epsilon_g \sigma T_g^4 + F_{12} J_2 \tau_{12} \qquad\qquad (5.103)$$

$$= 0.216 \times 5.67 \times 10^{-8} \times 1200^4 + 1 \times 37200 \times 0.784 \qquad (5.104)$$

$$= 54.4 \; \frac{kW}{m^2}$$

$$q_1 = J_1 - G_1 = 2.87 \times 10^4 - 54.4 \times 10^3$$

$$= 232 \; \frac{kW}{m^2} \qquad\qquad (5.105)$$

$$q_2 = J_2 - G_2 \qquad\qquad (5.106)$$

$$G_2 = \epsilon_g \sigma T_g^4 + F_{12} J_1 \tau_{12} \qquad\qquad (5.107)$$

$$= 25395.8 + 1 \times 2.87 \times 10^5 \times 0.784$$

$$= 2.5 \times 10^5 \; \frac{W}{m^2} \qquad\qquad (5.108)$$

$$q_2 = J_2 - G_2 = 37200 - 2.5 \times 10^5$$

$$= -2.13 \times 10^5 \; \frac{W}{m^2} \qquad\qquad (5.109)$$

If we are given a black enclosure or a problem involving two black surfaces, parallel plate formula, we can either use the solution and get the exponential integral or we can use the theory of evacuated enclosures and solve the problem.

As common sense will tell us, the most important point is that radiation is leaving the left wall and arriving at the right wall. Therefore, the net heat flux from the left wall will have to be positive because that is the wall at the highest temperature. The gas and this wall are at higher temperatures compared to the right side wall. The right side wall, being at a lower temperature, must receive radiation from both the left wall and the intervening gas. Therefore, intuitively one would expect q_2 to be negative. But the beauty is that $q_1 \neq q_2$.

This means that the situation is unbalanced. We do not have equilibrium here and, therefore, the radiative transfer equation is not an expression of the law of conservation of energy. At a particular instant of time, what is the net heat flux which is going out? This is the question it tries to answer. We have to combine it with some energy equation if we have to get the temperature distribution. Alternatively, we can pose the problem like this. For equilibrium to prevail between these two parallel plates, what should the gas temperature be? That means we will have to start with $q_1 = -q_2$, treat T_g as unknown and determine the resultant temperature, we do it in the next problem.

Example 5.4 Revisit the previous problem for the case of radiative equilibrium. Determine the gas temperature T_g. All other parameters are the same as before.

Solution: The solution to a radiative transfer equation does not guarantee equilibrium. The gas is getting heated in this case. The left wall is giving out 232 kW/m^2 while the right wall is getting only 213 kW/m^2. So 19 kW/m^2 of energy is being absorbed by the gas every second. The gas has to get heated up. So, the equilibrium temperature should be above 1200 K.

Radiative equilibrium case:

$$q_1 = J_1 - G_1 \tag{5.110}$$

$$q_2 = J_2 - G_2 \tag{5.111}$$

$$q_1 = -q_2 \text{(for radiative equilibrium)} \tag{5.112}$$

$$J_1 = \sigma T_1^4 = 2.87 \times 10^5 \frac{W^2}{m} \tag{5.113}$$

$$J_2 = \sigma T_2^4 = 37200 \frac{W^2}{m} \tag{5.114}$$

$$G_1 = \epsilon_g \sigma T_g^4 + F_{12} J_2 \tau_{12} \tag{5.115}$$

$$G_1 = 0.216 \times 5.67 \times 10^{-8} \times T_g^4 + 29164 \tag{5.116}$$

$$\tau_{12} = 0.784 \tag{5.117}$$

Fortunately, the mean beam length is not dependent on temperature. If it were so, we would get into a loop. We made a statement earlier that we are able to separate the geometric and thermal parts. Now, we can appreciate the significance of that statement. If the thermal and geometric parts were combined, it would lead to tedious iterations. Fortunately view factors are also not dependent on temperatures.

$$G_2 = \epsilon_g \sigma T_g^4 + F_{12} J_1 \tau_{12} \tag{5.118}$$

$$= 0.216 \times 5.67 \times 10^{-8} \times T_g^4 + 2.25 \times 10^5 \tag{5.119}$$

$$q_1 = -q_2 \tag{5.120}$$

$$\therefore J_1 - G_1 = -J_2 + G_2 \tag{5.121}$$

$$J_1 + J_2 = G_1 + G_2 \tag{5.122}$$

$$2.87 \times 10^5 + 37200 = 0.216 \times 5.67 \times 10^{-8} \times T_g^4 + 29164$$
$$+0.216 \times 5.67 \times 10^{-8} \times T_g^4 + 2.25 \times 10^5 \tag{5.123}$$

$$2.87 \times 10^5 + 37200 = 0.432 \times 5.67 \times 10^{-8} \times T_g^4 + 2.54 \times 10^5 \tag{5.124}$$

$$T_g^4 = 2.859 \times 10^{12} \tag{5.125}$$

$$T_g = 1300 \text{ K} \tag{5.126}$$

If we want the gas to be under radiative equilibrium, it should be at a temperature of 1300 K or left to itself, if sufficiently long time has elapsed, and we are not controlling the gas temperature, the gas will come to this temperature. Then, whatever is coming from the left side will go to the right side.

In the absence of the gas, what will q be? In this case, with the gas, $q_1 = -q_2 = 2.22 \times 10^5 \frac{W}{m^2}$ We want to appreciate the three cases:

• $T_g = 1200$ K, $q_1 = 2.32 \times 10^5 \frac{W}{m^2}$, $q_2 = -2.13 \times 10^5 \frac{W}{m^2}$. If all the temperatures are specified, we have no control over the energy balance.

- At radiative equilibrium: $T_g = 1300$ K, $q_1 = -q_2 = 2.22 \times 10^5 \frac{W}{m^2}$ We talk about energy balance in the equilibrium case when $q_1 = -q_2$ Even in this case, because the gas has a $\kappa \neq 0$, we are getting a flux lower than what we would have got if we had a transparent gas or vacuum between the two plates.
- Gas with $\kappa = 0$, $q_1 = -q_2 = 2.5 \times 10^5 \frac{W}{m^2}$

So by working out the last 3 problems, many of the concepts in gas radiation become clear. Because of the absorption of the gas, though it is emitting, the gas is at a temperature in between the other two walls and the net effect is that it retards the flow of heat from one wall to the other walls.

There could be another case where the gas is very hot and we want the heat to be transferred to the other walls or a flowing fluid. Such a situation arises in a fire tube boiler, where the gas is hot and the water gets this heat and becomes steam. Gas radiation plays a very important role in this process. Water flows outside of the tubes which becomes steam because of the heat from the gases.

In this case (i.e., Example 5.2), the walls are hot and the gas is getting heated up due to the presence of the wall. The gas will reduce the radiative heat transfer between the two walls, because of its absorptive characteristics.

5.5.4 Calculation of Emissivities and Absorptivities in Mixtures of Gases

Mixtures of gases are very important, as we find them in combustion chambers, furnaces and exhaust gases of automobiles. Wherever hydrocarbons are burnt, the resultant products contain water vapour and carbon dioxide. Both are radiatively participative and they interact such that they have overlapping bands which cannot be separated. Radiation entering Earth's outside atmosphere follows the Planck's distribution for a body at 5800 K while entering, but the same becomes zig-zag after getting attenuated by gases like water vapour and CO_2. If we want to do a detailed calculation, we have to solve the equation of transfer for every spectral band knowing the properties of absorptivity and emissivity which must come from molecular spectroscopy and then solve it band by band or interval by interval. This is called the **line by line model**, which is very advanced and time consuming.

However, for practical purposes, if we want to have a first cut analysis or design of a combustion chamber or a furnace and so on, we do not have to do those detailed calculations but opt for a simplified approach instead. We use some tables or charts, do minimal calculations and arrive at the required information. These charts were developed by Hottel, mainly for the $H_2O + CO_2$ mixture at 1 atm pressure.

The gas emissivity, ϵ_g should, in general, be a function of

$$\epsilon_g = f(L_m, P_g, P, C, T_g)$$
$$L_m = \text{mean beam length}$$
$$P_g = \text{partial pressure of the gas}$$

$$P = \text{total pressure of the gas}$$
$$C = \text{concentration of other gases}$$
$$T_g = \text{temperature of the gas}$$

The temperature is included here as we have already shown that emissivity is a function of temperature. We have included the concentration of other gases because there could be interaction and overlap of absorption bands. There could be bands in which more than one gas absorbs and so on. The total pressure affects the emissivity because it decides how much gas is present in a particular volume which decides the intermolecular spacing, which in turn affects its capacity to absorb and emit. Baseline charts have been developed for a pressure of 1 atmosphere and the correction charts help give the corrections for pressures other than 1 atmosphere.

Principally, we are considering two gases—$H_2O + CO_2$

- The emissivity of CO_2 is a function of the partial pressure of CO_2 multiplied by the mean beam length and its gas temperature.

$$\epsilon_C = f_1(P_C L_m, T_g)$$

The product of mean beam length and pressure is given in SI units as atm-m in the chart (see Fig. 5.10). The chart will directly give us the gas emittance and we can see that the maximum is 0.3. Suppose, we consider two parallel plates with $L = 1.5$ m, mean beam length $= 1.8 \times L$. If the total pressure is 2 bar and the mole fraction of CO_2 is 0.4, the partial pressure of CO_2 is 0.8 bar. Now if we calculate and get $P_C L_m$, we see in the chart that various curves are available for various values of $P_C L_m$. T_g is the gas temperature given in Kelvin. From this, we can straightaway read the value of ϵ_g. For pressures other than 1 atm as in the case that we just saw, we use the correction factor given in Fig. 5.11.
- Water vapour

$$\epsilon_w = f_2(P_w L_m, T_g)$$

Following a similar procedure, we will read the emissivity value of water vapour from the charts given in Fig. 5.12. If the $P_{\text{total}} \neq 1$ atm, we apply the correction factors from the charts again (Fig. 5.13).

$$c_c = f_3(P, P_c L_m) \qquad (5.127)$$
$$c_w = f_4\left(\frac{P_w + P}{2}, P_w L_m\right) \qquad (5.128)$$

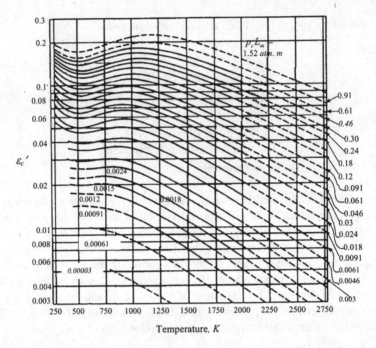

Fig. 5.10 Emissivity of carbon dioxide in a mixture with nonradiating gases at a total pressure of 1 atm and of hemispherical shape

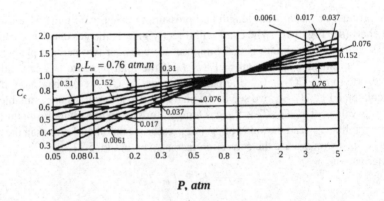

P, atm

Fig. 5.11 Correction factor for obtaining carbon dioxide emissivities at pressures other than 1 atm

These charts are mostly empirical, and results were obtained after many experiments were done with different concentrations of gases.

Then there are some spectral bands in which the absorption of carbon dioxide and water vapour overlap. So, we cannot simply add the emissivity of carbon dioxide with the emissivity of water vapour. If we want to get the total emissivity,

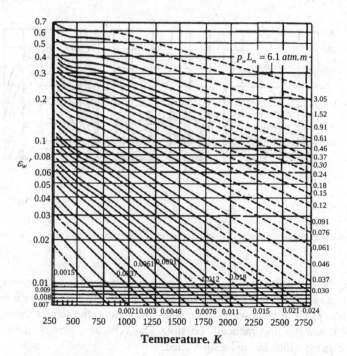

Fig. 5.12 Emissivity of water vapour in a mixture with nonradiating gases at a total pressure of 1 atm and of hemispherical shape

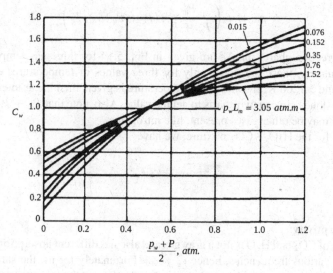

Fig. 5.13 Correction factor for obtaining water vapour emissivities at pressures other than 1 atm

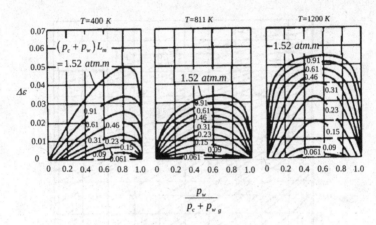

Fig. 5.14 Correction factor associated with mixtures of water vapour and carbon dioxide

we may be inclined to believe that it would be correct to do it as $(\epsilon_c + \epsilon_w)$. This should be valid for 1 atm pressure. For other pressures, we would say it should be $(c_c\epsilon_c + c_w\epsilon_w)$. But what will happen is that sometimes the sum of these two will exceed 1, which is not practically possible. That is because we have not taken care of the overlap which has to be subtracted.

Since the spectra of CO_2 and $H_2O + CO_2$ overlap, a correction is required.

$$\Delta\epsilon = f_5\left(\frac{P_w}{P_w + P_c}, (P_c + P_w)L_m, T_g\right) \qquad (5.129)$$

The correction charts for $\Delta\epsilon$ are given in Fig. 5.14 for three gas temperatures. Unfortunately, these are given only for three values of temperatures —400 K, 810 K and >1200 K. If some other temperature is given, then linear interpolation must be done between two charts to get the value. Also, note that $P_w + P_c \neq P_{total}$ as there may be other gases present, like nitrogen.

Finally, for the $H_2O + CO_2$ mixture, we have

$$\epsilon_g = c_c\epsilon_c + c_w\epsilon_w - \Delta\epsilon \qquad (5.130)$$

Gas absorptivity:

A mixture of CO_2 and H_2O is not a gray gas and also has different absorption spectral bands for various frequencies, hence $\epsilon_g \neq \alpha_g$. Fortunately for us, the same charts can be used for α_g also too, but with some changes. For ϵ_g, the gas temperature (T_g) is very important. On the other hand, for α_g, the surface temperature (T_s) will be very important.

$$\alpha_c = c_c \epsilon_c \left(\frac{T_g}{T_s} \right)^{0.65} \qquad (5.131)$$

$$\alpha_w = c_w \epsilon_w \left(\frac{T_g}{T_s} \right)^{0.45} \qquad (5.132)$$

$$\text{where now,} \quad \epsilon_c = f_6 \left(P_c L_m \frac{T_s}{T_g}, T_s \right) \qquad (5.133)$$

$$\epsilon_w = f_7 \left(P_w L_m \frac{T_s}{T_g}, T_s \right) \qquad (5.134)$$

In the above equation, T_s is surface temperature. A correction to take care of overlap is given below

$$\Delta\alpha = f_8 \left(\frac{P_w}{P_w + P_c} \cdot \frac{T_s}{T_g}, (P_c + P_w) L_m \frac{T_s}{T_g}, T_s \right) \qquad (5.135)$$

$$\boxed{\alpha = \alpha_c + \alpha_w - \Delta\alpha} \qquad (5.136)$$

As an alternative to the tedium of using these charts, one can use detailed gas emissivity relations developed by Prof.Leckner and his group at Chalmers University, Sweden. Interested readers may refer to Leckner (1972)

Example 5.5 A furnace having a spherical cavity of 0.5 m diameter contains a gas mixture at 2 atm and 1400 K. The mixture consists of CO_2 at a partial pressure of 0.6 atm, N_2 with a partial pressure of 0.9 atm and the remaining is water vapour. If the cavity wall is black, what is the cooling rate required to maintain its temperature at 500 K?

Solution:

$$L_m = \text{Mean beam length} = 0.65D = 0.325 \text{ m} \qquad (5.137)$$

$$P_c L_m = 0.6 \times 0.325 = 0.195 \text{ atm m} \qquad (5.138)$$

$$\text{From the charts, } \epsilon_c = 0.13 \qquad (5.139)$$

$$\text{Correction factor for carbon dioxide } c_c = 1.2 \qquad (5.140)$$

$$\text{Correction factor for water vapour } c_w = 1.57 \qquad (5.141)$$

$$P_w L_m = 0.5 \times 0.325 = 0.163 \text{ atm m} \qquad (5.142)$$

$$\text{From the charts, } \epsilon_w = 0.14 \qquad (5.143)$$

$$(P_c + P_w) L_m = 1.1 \times 0.325 = 0.35 \text{ atm m} \qquad (5.144)$$

$$\frac{P_w}{P_w + P_c} = \frac{0.5}{1.1} = 0.45 \qquad (5.145)$$

$$\Delta\epsilon(T_g = 1400 \text{ K}) = 0.03 \qquad (5.146)$$

$$\epsilon_g = c_c \epsilon_c + c_w \epsilon_w - \Delta\epsilon \tag{5.147}$$
$$\epsilon_g = 0.13 \times 1.2 + 0.14 \times 1.57 - 0.03 \tag{5.148}$$
$$\epsilon_g = 0.3458 \tag{5.149}$$

We repeat the procedure for calculation of α_g. However, we need to remember that we have to apply the correction in this case as

$$\alpha_c \text{ (from the charts)} = 0.09 \tag{5.150}$$
$$\alpha_w \text{ (from the charts)} = 0.17 \tag{5.151}$$
$$c_c = 1.2 \tag{5.152}$$
$$c_w = 1.57 \tag{5.153}$$
$$\alpha_c = c_c \alpha_c \left(\frac{T_g}{T_s}\right)^{0.65} = 0.211 \tag{5.154}$$
$$\alpha_w = c_w \alpha_w \left(\frac{T_g}{T_s}\right)^{0.45} = 0.4242 \tag{5.155}$$
$$\Delta\alpha = f_8\left(\frac{P_w}{P_w + P_c}, (P_c + P_w)L_m, T_s\right) \tag{5.156}$$
$$\Delta\alpha = 0.005 \tag{5.157}$$
$$\alpha_g = \alpha_c + \alpha_w - \Delta\alpha = 0.6302 \tag{5.158}$$

The most important thing we see here is that $\epsilon_g \neq \alpha_g$. A mixture of carbon dioxide and water vapour is a non-gray gas.

$$q_1 = J_1 - G_1 \tag{5.159}$$
$$Q_1 = (J_1 - G_1)4\pi R^2 \tag{5.160}$$
$$\epsilon_1 = 1 \tag{5.161}$$
$$J_1 = \epsilon_1 \sigma T_1^4 + (1 - \epsilon_1) = \sigma T_1^4 \tag{5.162}$$

$$G_1 = \epsilon_g \sigma T_g^4 + \tau_g(\sigma T_1^4) \tag{5.163}$$
$$\alpha_g + \tau_g + \rho_g = 1; \ \rho_g = 0 \tag{5.164}$$
$$\text{(In view of the fact that we have neglected} \tag{5.165}$$
$$\text{scattering both from and to the gas).} \tag{5.166}$$
$$\alpha_g + \tau_g = 1; \ \tau_g = 1 - \alpha_g \tag{5.167}$$
$$G_1 = \epsilon_g \sigma T_g^4 + (1 - \alpha_g)(\sigma T_1^4) \tag{5.168}$$
$$q_1 = J_1 - G_1 = \sigma T_1^4 - \epsilon_g \sigma T_g^4 - \sigma T_1^4 + \alpha_g \sigma T_1^4 \tag{5.169}$$
$$q_1 = \alpha_g \sigma T_1^4 - \epsilon_g \sigma T_g^4 \tag{5.170}$$

For this problem under consideration,

$$q_1 = 0.64 \times 5.67 \times 10^{-8} \times 500^4$$
$$- 0.36 \times 5.67 \times 10^{-8} \times 1500^4 \tag{5.171}$$

$$q_1 = -101.7 \, \frac{kW}{m^2} \tag{5.172}$$

$$Q_1 = q_1 4\pi R^2 = -79.378 \, kW \tag{5.173}$$

The solution tells us that we need to cool the cavity, which is very evident. Most importantly, in this spherical cavity, if we have ethane, propane, butane or LPG, burn it, allow it to reach a temperature of 1500 K and we get a mixture of $CO_2 + H_2O$, it is possible to have tubes on the outside of the spherical cavity, send water through them which will get heated at the rate of 60 kW. This is the radiative heat transfer power that the gas is capable of transferring to the wall. This will be very important in furnace calculations, design of radiant super heaters and so on. Needless to say, these ideas have several engineering applications!

Problems

5.1 Two very long parallel plates are maintained at uniform temperatures of $T_1 = 850$ K and $T_2 = 550$ K. The respective emissivities are 0.6 and 0.8. Between the two plates, an absorbing and emitting gas at a uniform temperature of $T_g = 400$ K flows with an absorption coefficient of 0.07 m^{-1}. The spacing between the plates, L is 1.2 m. Neglecting any convection between the gas and the plates, compute the net radiative heat flux at the two walls in kW/m^2.

5.2 An absorbing and emitting gas at a temperature of $T_g = 1100$ K flows between two very long parallel plates. Both the plates have an emissivity of 0.8 and are maintained at a uniform temperature of $T_1 = T_2 = 500$ K. The spacing between the plates, L is 1 m. Determine the absorption coefficient of the gas if the amount of cooling required at each wall surface is $q = 35$ kW/m^2 (Please note that this is a classic inverse problem where one or more properties are to be estimated from measured heat fluxes or temperatures, as the case may be).

5.3 A cubical furnace 1 m on each side is made up of a mixture of 40% CO_2 and 50% H_2O at a total pressure of 1 atm. The remainder is Nitrogen. The gas temperature is uniform at 1900 K and the walls are maintained at 1000 K. The inner surfaces of the furnace are black. Determine the total heat removed from the walls in order to maintain the temperatures.

5.4 Repeat Problem 5.3 for the case of the wall emissivities being 0.8. Comment on your results.

5.5 A rectangular furnace is of dimensions $0.45 \times 0.65 \times 2.0$ (all in m). The interior walls have a hemispherical total emissivity of 0.8 and are maintained at 750 K. The furnace is filled with combustion products at a temperature of 2100 K. The composition of the combustion products by volume is 42% of CO_2, 22% of water vapour and the remainder N_2. The total pressure is 2.5 atm. Calculate the net radiative heat flux to the walls of the furnace using the charts for gas emissivity of mixtures and the radiosity method applied to enclosures, modified for participating media.

Chapter 6
Introduction to Atmospheric Radiation

6.1 Introduction

In this chapter, the basics of radiative transfer in planetary atmosphere are introduced. First, we look at radiation spectra, followed by black body radiation for temperatures commonly encountered in planetary atmosphere. This is followed by a consideration of RTE for a plane parallel atmosphere that is emitting and absorbing. We then discuss the case of radiative equilibrium followed by a brief description of infrared remote sensing.

6.2 Electromagnetic Spectrum

As already discussed in earlier chapters, the speed of light (c), wavelength (λ) and frequency (ν) are related by

$$\nu = \frac{c}{\lambda} \tag{6.1}$$

In atmospheric science, $\lambda < 4\,\mu m$ refers to shortwave radiation and is typically associated with solar radiation. Long wave radiation ($\lambda > 4\,\mu m$) is typically associated with terrestrial radiation from the earth. As already mentioned, visible radiation has a wavelength range of $0.4 - 0.7\,\mu m$. Radiation with wavelength less than 0.4 and greater than 0.01 μm is termed as ultra violet radiation. The range $0.7 - 4\,\mu m$ is known as near infrared (IR) and $4 - 100\,\mu m$ is known as far infrared. Radiation in the range $1 - 10$ mm is known as microwave radiation. The energy balance of the earth is largely decided by the incoming solar radiation which is mostly in the UV, visible and near infrared radiation and the earth's radiation itself which peaks around 10 mm (Far infrared). Microwave radiation has no role in this. However, microwave radiation from the earth can penetrate clouds and because of this reason is extremely useful in both passive and active remote sensing. It is instructive to mention that based on the relation $E = h\nu$, the energy associated with microwave radiation emission is

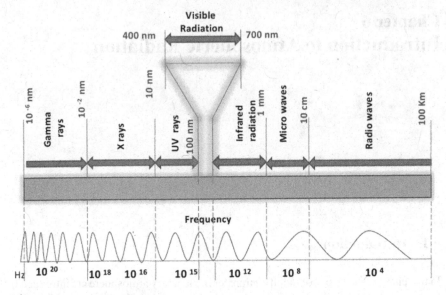

Fig. 6.1 Electromagnetic spectrum

small, and hence, remote sensing of the atmosphere with space-borne instruments to detect the radiation has to be necessarily done with low earth-orbiting satellites. Figure 6.1 shows a simplified representation of the electromagnetic spectrum.

6.3 Black Body Radiation

The basic radiation laws namely Planck's distribution, Wien's law and Stefan Boltzmann can all be applied in atmospheric radiation with the assumption that the sun is a black body. The Stefan–Boltzmann law can also be applied to determine the Earth's equivalent black body temperature, as will be shown shortly.

6.3.1 Temperature of the Sun

Example 6.1 Taking the mean sun-earth distance to be 1.49×10^{11} m and the radius of the outermost visible layer of the sun R_{sun} to be 6.96×10^{8} m, determine the equivalent black body temperature T_{eq} of the sun (i.e sun's outermost visible layer also known as solar photosphere) if the intensity of solar radiation reaching the earth is 1353 W/m^2.

Solution:

$$q_{\text{photosphere}} = 1353 \left[\frac{1.49 \times 10^{11}}{6.96 \times 10^8} \right]^2 \tag{6.2}$$

(We can use the inverse square law in the above expression)

$$q_{\text{photosphere}} = 6.2 \times 10^7 \, \frac{W}{m^2} \tag{6.3}$$

(The quantity $1353 \, \frac{W}{m^2}$ is known as the solar constant and is experimentally measured). Using Stefan–Boltzmann law, we have

$$\sigma T_{\text{eq}}^4 = 6.2 \times 10^7 \tag{6.4}$$

$$T_{\text{eq}} = \left[\frac{6.2 \times 10^7}{5.67 \times 10^{-8}} \right]^{0.25} \tag{6.5}$$

$$T_{\text{eq}} = 5751 \text{ K} \tag{6.6}$$

Example 6.2 From the equivalent black body temperature of the sun determined in the previous example and Wien's displacement law, determine the wavelength corresponding to the maximum intensity of solar radiation.

Solution:

$$\lambda_{\text{max}} T = 2898 \, \mu\text{mK} \tag{6.7}$$

$$\lambda_{\text{max}} = \frac{2898}{5751} = 0.504 \, \mu\text{m} \tag{6.8}$$

It is seen that λ_{max} is very much in the visible part of the EM spectrum.

6.3.2 Temperature of the Earth

The earth is not a perfect black body and the fraction of incoming radiation that is reflected, reflectivity, is around 0.3. In the parlance of atmospheric science, this reflectivity is known as **planetary albedo**.

Example 6.3 If the earth's albedo is 0.3 determine the equivalent black body temperature of the earth, assuming it to be in radiative equilibrium. The solar constant can be assumed to be 1353 W/m^2.

Solution:

$$4\pi R_E^2 \sigma T_{eq}^4 = \pi R_E^2 (1 - 0.3) \times 1353 \tag{6.9}$$

$$T_{eq} = \left[\frac{0.7 \times 1353}{4 \times 5.67 \times 10^{-8}} \right]^{0.25} \tag{6.10}$$

$$T_{eq} = 254 \text{ K} \tag{6.11}$$

6.4   Radiative Transfer Equation for a Plane Parallel Atmosphere

From the previous chapter, the RTE for a plane parallel atmosphere can be written as

$$\frac{dI}{ds} + kI = k\frac{\sigma T^4}{\pi} \tag{6.12}$$

where T is the temperature of the air layer in k, I is the intensity in $\frac{W}{m^2 sr}$ and k is the absorption coefficient. The use of Stefan–Boltzmann law in Eq. (6.12) tacitly implies that local thermal equilibrium (LTE) prevails in the atmosphere. However, this is true only for the lower atmosphere.

The plane parallel assumption reduces the problem to a one dimensional one. Now, we have only two fluxes I^+ and I^- in the upward and downward directions. $I^+ = \iint I(\theta) . \cos\theta \sin\theta d\theta d\phi$ integrated over the downward facing atmosphere and $I^- = \iint I(\theta) . \cos\theta \sin\theta d\theta d\phi$ integrated over the upward facing atmosphere.

6.5   Radiative Transfer Equation (RTE) for an Absorbing and Emitting Atmosphere

The RTE equation for monochromatic or spectral radiation intensity, I_λ (or I), as derived in the last chapter is,

$$\frac{dI_\lambda}{ds} + k_\lambda . I_\lambda = k_\lambda I_{b,\lambda}(T_g) \tag{6.13}$$

In Eq. (6.13), k_λ is the spectral absorptivity in m^{-1}, "s" is the direction under consideration. The first term on the left hand side represents the change in intensity, the second represents the attenuation by absorption and the right hand side represents augmentation by emission. Please note that if scattering by particles needs to be accounted, as is the case in microwave remote sensing, a source term needs to be added to the right hand side of Eq. (6.13). Furthermore, please also note that

Eq. (6.13) assumes a gray atmosphere consequent upon the appearance of k_λ in the emission term, which in turn arises from Kirchoff's law for a gray medium wherein $\epsilon_\lambda = \kappa_\lambda$. Equation (6.13) is frequently referred to as **Schwartzchild's** equation. Under the assumption of local thermodynamic equilibrium(LTE), $I_{b\lambda}$ can be replaced by Planck's law. The LTE does not hold good for the upper parts of the atmosphere, where the latter is rarefied.

In consonance with the terminology commonly used in atmospheric sciences, we will denote $I_{b,\lambda}$ as $B_\lambda(T)$. With this, Eq. (6.13) becomes

$$\frac{dI_\lambda}{ds} + k_\lambda.I_\lambda = k_\lambda B_\lambda(T) \tag{6.14}$$

Equation (6.14) is a first order linear differential equation, provided the source function (RHS) is known a priori, which is the case for an absorbing and emitting medium and for simplified cases of scattering. For atmospheric scattering, the solution to RTE is formidable. Multiplying both sides of Eq. (6.13) by $e^{k_\lambda s}$

$$e^{k_\lambda s}\left[\frac{dI_\lambda}{ds} + k_\lambda.I_\lambda\right] = e^{k_\lambda s}k_\lambda B_\lambda(T) \tag{6.15}$$

The LHS is actually $\frac{d(e^{k_\lambda s}.I_\lambda)}{ds}$; substituting for the LHS in Eq. (6.15)

$$\frac{d(e^{k_\lambda s}.I_\lambda)}{ds} = e^{k_\lambda s}k_\lambda B_\lambda(T) \tag{6.16}$$

Recall that $e^{k_\lambda s}$ is a factor. Integrating Eq. (6.16) from 0 to any s_1 along s, we have

$$\left(e^{k_\lambda s}.I_\lambda\right)_0^{s_1} = \int_0^{s_1} e^{k_\lambda s}.k_\lambda.B_\lambda(T)ds \tag{6.17}$$

$$\left(e^{k_\lambda s_1}.I_\lambda(s_1)\right) - I_\lambda(0) = \int_0^{s_1} e^{k_\lambda s}.k_\lambda.B_\lambda(T)ds \tag{6.18}$$

$$I_\lambda(s_1) = I_\lambda(0).e^{-k_\lambda s} + \int_0^{s_1} e^{k_\lambda(s-s_1)}.k_\lambda.B_\lambda(T)ds \tag{6.19}$$

In Eq. (6.19) $I_{\lambda(0)}$ at $s = 0$ is known. The first term on the right hand side of Eq. (6.19) represents the radiation intensity at $I_\lambda(0)$ arriving at s_1 after getting attenuated because of a non-zero k_λ. The second term represents augmentation of the radiation intensity because of emission (in fact $B_\lambda(T)$ can be more generally written as a source term S_λ in which case, the second term on the RHS of Eq. (6.19) represents augmentation by both emission and in scattering in the direction s). Equation (6.19) is an integral equation and as aforementioned obtaining I_λ for an anisotropic scattering can be challenging and demands the use of numerical techniques.

It is instructive to note that for $B_\lambda(T) \approx 0$, Eq. (6.19) reduces to

$$I_\lambda(s_1) = I_\lambda(0).e^{-k_{\lambda s}} \tag{6.20}$$

Equation (6.20) is the familiar Beer or Beer–Lambert law of radiation applicable for a strongly absorbing and weakly emitting gas. The plane parallel atmosphere considerably simplifies radiation calculations and the temperature, densities of atmospheric quantities are assumed to be function of only the height (or pressure). By introducing a new property called optical depth (τ) given by $k_\lambda.x$, the upward/downward spectral fluxes can be computed as follows

$$q_\lambda^\pm(\tau_\lambda) = \int\limits_0^{2\pi} \int\limits_0^{\pi/2} I_\lambda^\pm(\tau_\lambda \cos\theta) \cos\theta \sin\theta \, d\theta d\phi \tag{6.21}$$

6.6 Infrared Remote Sensing

To calculate the heat flux q^+ or q^- from the spectral fluxes q_λ^+ or q_λ^-, one has to integrate Eq. (6.21) over the desired wavelength interval $\lambda_1 - \lambda_2$ (or wave number). In order to be able to do this, the variation of τ_λ with λ needs to be known. Hence, the flux determination will involve a convolution of q_λ^+ and Planck's distribution, since q_λ usually varies rapidly with λ. Integration is usually done over a narrow range of wavelengths (or wave numbers), leading to what are known as line-by-line calculations, wherein typically for an IR instrument, millions of line-by-line calculations with available spectroscopic absorption data are to be done. This is a big challenge in passive remote sensing of the atmosphere which is done with the help of a multispectral instrument usually known as sounder. Repeated calculations with assumed profiles of the atmosphere become inevitable to solve the inverse problem of retrieving or estimating the atmospheric temperature and humidity profile from satellite flux densities (or radiances). A fast radiative transfer (RT) model is invariably required by satellite meteorologists. Though fast models are usually regression based, researchers have also begun using state-of-art simulation tools like Artificial Neural Network (ANN) to develop "fast RT models".

A fast RT model, by definition, is one which takes in the atmospheric profile, typically temperature and humidity and returns the fluxes within a time, orders of magnitude lower than a regular RTE solution. A fast RT model is built on a database of profiles vs fluxes, developed by repeated solutions to the RTE. They are trained and tested rigorously before they can be employed operationally. In actual remote sensing, the sensor characteristics, also known as the spectral response function (SRF), has to be convolved with Planck's distribution together with the variance of τ_λ vs λ to compute the radiances. For satellite meteorologists, knowledge of SRF's is required before atmospheric profiles of temperature and humidity can be retrieved from measured satellite radiances at designated frequencies. In a typical multispectral instrument usually 15–20 channels are present, while in a hyperspectral sounder thousands will be present. These are carefully chosen based on atmospheric absorption and transmission windows and the mission objectives. Readers are advised to look at advanced texts on atmospheric radiation or remote sensing to know more about the state-of-the-art in this field.

Problems

6.1 The mean sun-earth distance is 1.48×10^{11} m and the radius of the photosphere of the sun is 6.95×10^8 m. The equivalent black body temperature of the sun is 5800 K.

(a) With the above data, determine the intensity of the solar radiation reaching the earth (also known as the solar constant).

(b) The mean sun-earth distance given above is known as one astronomical unit (AU). If the Venus-Sun distance is 0.72 AU, determine the solar constant for Venus and compare it with the result obtained in part (a).

(c) Using the result obtained in part (b), determine the equivalent black body temperature of Venus, if the planetary albedo (reflectivity) for Venus is 0.77.

Chapter 7
Inverse Problems in Radiation

7.1 Introduction

Consider the case of a person with excruciating chest pain who is being wheeled into the emergency care unit of a hospital. Upon checking the patient's vitals and stabilizing him, if required, the intensivist and other doctors try to ascertain the cause of the pain. An ECG will be taken to rule out a myocardial infarction (heart attack) followed by a battery of tests. The exact cause of the pain could range from a simple indigestion related gas pain to a life threatening heart attack or even a cardiac arrest. In the language of engineering, the chest pain is our "measurement" or "data" with which its cause has to be identified. Needless to say the pain here is the effect. Since, there can be several causes to this pain, the problem at hand is challenging and ill posed, as several causes could lead to the same effect (pain). The goal, then, is to identify the correct cause. This example is a classic case of an inverse problem. An important point to be noted in the foregoing example is the importance of the physician's skills in quickly and deftly sifting through the symptoms the patient presents, results of the examinations and tests and how he/she correlates here with similar cases seen in the past. His "expert" knowledge is a distinguishing feature that holds the key to tackle the ill-posedness. Much in the same way, the prior knowledge of the problem goes a long way in reducing the ill-posedness associated with most problems in science and engineering. An analyst who uses such prior knowledge for better estimation of parameters or causes is frequently referred as a Bayesian.

Similarly, in sciences, in general, and radiation, in particular, there are situations where one needs to identify or establish the correct cause or the set of causes from the effect(s). In radiation, more often than not the effect is temperature time history, temperature distribution or heat flux distribution. The cause we are seeking could be a thermophysical property like emissivity, thermal conductivity or thermal diffusivity in a multimode problem involving radiation with conduction and/or convection. Sometimes one may be interested in obtaining the estimate of the heat transfer or mass transfer coefficient, which are known as transport properties in a combined problem or in a problem involving only convection. In quite a few engineering applications, the goal can also be the estimation of the heat flux or the heat flux distribution, which presents itself, invariably, as a boundary condition.

A classic, frequently quoted example in inverse heat transfer is the problem of determination of surface heat flux in a re-entry vehicle. When such a vehicle re-enters the earth's atmosphere from outer space the velocities encountered are enormous as, for example, 8 km/s (translating to a Mach number of the order of 25) leading to a massive aerodynamic heating at the surface, wherein the kinetic energy of the air is converted to enthalpy rise as a result of the "braking" action on the fluid. Very high temperatures on the surface forbid us from placing heat flux gauges on the vehicle surface. Even so, however, the surface heat flux is a critical design parameter which needs to be known, in order to design, among other things, the thermal protection system. In view of this, temperature measurements are made at convenient locations on the inside of the vehicle with the help of thermocouples. Using the "measurements" as data, an inverse heat transfer model is set up, wherein guess values of the surface heat flux are given and temperatures corresponding to the measurement locations are computed by solving the appropriate governing equations for the problem under consideration. The sum of the squares of the deviations between measurements and simulations is usually minimized to obtain what is frequently referred to as the maximum likelihood estimate of the parameters.

Examples of inverse problems in radiation:

- Estimation of radiative surface properties like emissivity and absorptivity.
- Estimation of the absorption coefficient in tissues with help of CT or CAT (Computerized Axial Tomography) scans. The absorption coefficient can reveal information on whether an underlying pathology(i.e. disease) is present.
- Remote measurement of global rainfall through infrared or microwave combined sensors placed in satellites.

7.2   Least Squares Minimization for Parameter Estimation

In the previous section, an introduction to inverse problems in thermal sciences, in general, and radiation heat transfer, in particular, was presented. Atmospheric remote sensing is one field whose progress critically hinges on our ability to invert satellite measured quantities into geophysical parameters. There are many such fields where inverse problems are and will become important. Parameter estimation problems are invariably posed as optimization problems. Oftentimes, minimization of the sum of the least squares of the residue is done. Mathematically, if $Y_{\text{data},i}$ is the measured data vector and $Y_{\text{sim},i}$ is the simulated or calculated vector of Y values for assumed values of the parameters, then, we define the sum of the squares of the residue S(X) as

$$S(X) = \sum_{i=1}^{N}(Y_{\text{data},i} - Y_{\text{sim},i})^2 \tag{7.1}$$

Where N refers to the total number of measurements. The goal then is to minimize S(X) and X refers to the set of parameters that need to be estimated.

If each measurement is associated with a different error, given by a standard deviation σ_i, Eq. (7.1) can be modified as

$$S(X) = \sum_{i=1}^{N} \frac{(Y_{\text{data},i} - Y_{\text{sim},i})^2}{\sigma_i^2} \tag{7.2}$$

Minimization of S(X) in Eq. (7.2) is known as weighted least squares minimization. A different σ for each measurement makes the weighted least squares more general than the plain least squares minimization.

Example 7.1 Consider one-dimensional steady conduction in an infinitely long slab of thickness 100 mm made of insulating material whose thermal conductivity is k = 1 W/mK (Fig. 7.1). The left end of the slab is maintained at a temperature of 100 °C while the right end is exposed to the evacuated environment at 30 °C. There is no heat generation within the slab. The temperature distribution measured at eight locations across the slab is given below (Table 7.1). Using the principle of least square minimization, determine the hemispherical total emissivity of the exposed surface of the slab at its right end. Use finite differences to solve for the temperature

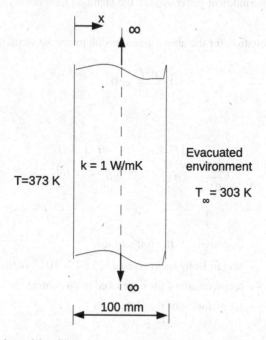

Fig. 7.1 Infinitely long slab subjected to a constant temperature at one end and radiating to the evacuated environment at the other end

Table 7.1 Temperature versus distance, for Example 7.1

x, mm	10	20	30	40	50	60	70	80
T, K	371.2	370.4	367.8	366.2	363.3	362	361.3	357.9

distribution in the slab for guess values of emissivity in the range of $0.15 \leq \varepsilon \leq 0.95$, with a search interval of 0.1.

Solution:

As is evident, this is typically an inverse problem since it requires the estimation of a transport property like emissivity from measured temperature distribution. Generally, the direct problem for such a case would be the determination of temperature distribution for a given set of thermophysical properties. There is only one, namely, emissivity in this case. The assumptions pertinent to the present problem are listed below:

1. The slab is infinitely long and finitely thick, facilitating one dimension heat conduction along only the axial direction.
2. The thermophysical properties for the slab are constant and do not change either with temperature or location.
3. The right wall loses heat to a constant temperature evacuated space only by radiation and convection currents are completely absent.
4. There is no internal heat generation in the slab and temperatures are recorded at steady state.

The governing equation for the above-stated problem can be written as

$$\frac{d^2 T}{dx^2} = 0 \qquad (7.3)$$

subject to

$$T = 373 \text{ K} \qquad at \ x = 0 \qquad (7.4)$$

$$-k \frac{dT}{dx} = \varepsilon \sigma (T^4 - T_\infty^4) \ at \ x = 100 \text{ mm} \qquad (7.5)$$

Where

$\varepsilon \rightarrow$ emissivity of the foil surface

$\sigma \rightarrow$ Stefan Boltzmann constant, $5.67 \times 10^{-8} \text{ W/m}^2 K^4$

$T_\infty \rightarrow$ temperature of the evacuated environment, K

$T \rightarrow$ temperature of the slab, K

Solution to the forward model:

The governing equation as given by Eq. (7.3) along with the boundary conditions as specified in Eqs. (7.4) and (7.5) can be discretized using second order finite differences. This would lead to a set of simultaneous linear equations in T, which then can be solved using an iterative method such as the Gauss Siedel scheme. A sample FORTRAN program for the problem is listed hereby for ready reference.

```fortran
1    program forward_model
2        implicit none
3    INTEGER::I
4    Integer,parameter:: Nx=11   ! No. of grid points in x direction
5    double precision :: T0(Nx),T(Nx),ITER, ERR(Nx), ermax,EPS,tol,
         lx=0.10,dx,E=0.83,a
6    double precision :: k=1,Tamb=303,sigma=5.67e-8,dummy
7
8    !T0 : Represents temperature in previous iteration .
9    !T  : Represents temperature in current iteration
10   !ITER : Iteration number
11   !ERR: error
12   !ermax: maximum error
13   !EPS: maximum permissible error
14   !tol: tolerance
15   !lx :length of computational domain in x direction
16   !dx: length of control volume
17   !E: emissivity
18
19   T(:)=0
20
21   ITER=0
22
23   EPS = 1e-8 !maximum error permissible or convergence crieterion
24
25   dx=lx/(Nx-1)
26
27   a=-k/(dx)
28
29   tol=1
30
31   !starting iterations for solving Laplace equation(Gauss Seidel
         Solver)
32
33   DO WHILE(tol> EPS)
34
35   ITER=ITER+1
36
37   T0(:) = T(:)
38
39   DO I=2,Nx-1
40
41   T(I)= (T(I-1)+T(I+1))/2
42
43   END DO
```

```fortran
44
45   !Boundary conditions
46
47
48   T(1)=373
49
50   T(Nx)=E*SIGMA*((T0(Nx))**4-(Tamb**4))+a*(4*T(Nx-1)-T(Nx-2))
51
52   T(Nx)=T(Nx)/(3*a)
53
54   ermax = 0.0
55
56   DO I=1,Nx
57
58   ERR(I) = T(I) - T0(I)
59
60   if(ABS(ERR(I)) > ermax)THEN
61
62      ermax =ABS( ERR(I))
63
64    ENDIF
65
66   END DO
67
68
69   write(*,*) "Residual = ",ermax
70
71   if(ermax < EPS)THEN
72
73   WRITE(*,*)"Solution converged in",ITER,"iterations"
74
75      ENDIF
76
77   tol = ermax
78   END DO
79   OPEN(unit=5,file="temperature.txt",status="unknown")
80
81   DO i=1,nx
82
83   WRITE(5,*)T(i)
84
85   END DO
86
87   CLOSE(5)
88
89   end program forward_model
```

Table 7.2 Variation of the sum of the residues, $S(\varepsilon)$ with emissivity(ε) for Example 7.1

ε	0.15	0.25	0.35	0.45	0.55	0.65	0.75	0.85	0.95
$S(\varepsilon)$	386.02	259.05	169.75	98.84	52.84	23.57	7.81	3.01	7.14

Fig. 7.2 Variation of residuals with emissivity for Example 7.1

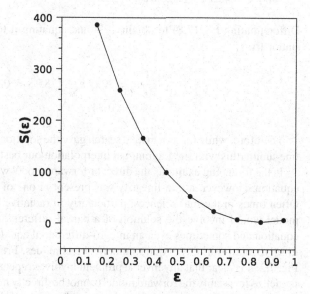

Least squares minimization can be used to solve the inverse problem by substituting different values of ε and solving the governing equation to obtain the temperature distribution T_i, where the subscript i denotes the location at which the temperature is measured. Once this is obtained, the sum of the residues is calculated as follows:

$$S(\varepsilon) = \sum_{i=1}^{N}(T_{\exp,i} - T_{\text{calc},i})^2 \qquad (7.6)$$

The sum of the square of the residual, $S(\varepsilon)$ is computed for ε ranging from $0.15 \leq \varepsilon \leq 0.95$ in steps of 0.1 and these are as shown in Table 7.2. Figure 7.2 depicts the plot of residuals $S(\varepsilon)$ for different values of ε. From Table 7.2 and Fig. 7.2, one can conclude that ε should lie in the range $0.75 \leq \varepsilon \leq 0.95$.

The exercise we have performed here is frequently referred to as exhaustive equal interval search. It is possible to employ faster and more sophisticated search algorithms like the Fibonacci or Golden Section search to get a much better estimate of ε with the same number of functional evaluations (Please refer to Balaji 2019 for further discussion of different optimization methods for a single and multivariable problem). Even with the slightly crude exhaustive search method presented above, we can fit a local Lagrangian interpolating polynomial for $S(\varepsilon)$ by using three values of ε where the residuals approach minimum, i.e. in the range $0.75 \leq \varepsilon \leq 0.95$.

$$S(\varepsilon) = \frac{(\varepsilon - 0.85)(\varepsilon - 0.95)}{(0.75 - 0.85)(0.75 - 0.95)} \times 7.81 + \frac{(\varepsilon - 0.75)(\varepsilon - 0.95)}{(0.85 - 0.75)(0.85 - 0.95)} \times 3.01$$

$$+ \frac{(\varepsilon - 0.75)(\varepsilon - 0.85)}{(0.95 - 0.75)(0.95 - 0.85)} \times 7.14 \tag{7.7}$$

$$S(\varepsilon) = 1048.5 \, \varepsilon^2 - 1785.8 \, \varepsilon + 757.38 \tag{7.8}$$

Differentiating Eq. (7.8) to obtain $\frac{dS(\varepsilon)}{d\varepsilon}$ and equating it to zero we can make $S(\varepsilon)$ stationary

$$\frac{dS(\varepsilon)}{d\varepsilon} = 2097 \, \varepsilon - 1785.8 = 0 \tag{7.9}$$

$$\varepsilon = 0.84 \tag{7.10}$$

Therefore, while the exhaustive search gave the solution as $0.75 \leq \varepsilon \leq 0.95$, upon fine-tuning this with the Lagrangian interpolation our best estimate of ε is 0.84.

In the foregoing example, the direct or forward model was an ordinary differential equation. However, a non-linearity was present in one of the boundary conditions. Often times in thermal sciences, particularly in radiative heat transfer, the forward model would involve the solution of a partial differential equation or an integral equation and sometimes even an integro-differential equation. These are formidable to solve and invariably require numerical techniques. From the flow chart given in Fig. 7.3, it is clear that any inverse problem involves repeated solution to the forward model, as invariably the forward model cannot be directly inverted. For example, there is no way by which we can write out a closed form expression for ε in Example 7.1 in term of temperatures and determine it right away. This is further compounded by the presence of noise in the measurement of temperature. The requirement of a repeated solution to the forward model often makes the solution of an inverse problem time-consuming. Researchers have tried to address this by developing a faster equivalent of the forward model using techniques like artificial neural networks, which are known as surrogates. From the flow chart, it is also evident that the solution to the inverse problem can often be posed as a minimization problem and so advanced and cutting edge optimization techniques can be made use of to improve accuracy, speed and robustness of the ever-increasing challenge posed by the high dimensional inverse problems, wherein several parameters need to be estimated from limited measurements.

7.3   The Bayesian Method for Inverse Problems

In the previous section, the basic concepts involved in solving an inverse problem were presented. A specific case of single parameter estimation in a combined conduction radiation problem was presented. It is now fairly straightforward for us to see that in a multiparameter problem, several combinations of parameters or the "causes" may lead to the same "effect". In view of this, an inverse problem is essentially ill-

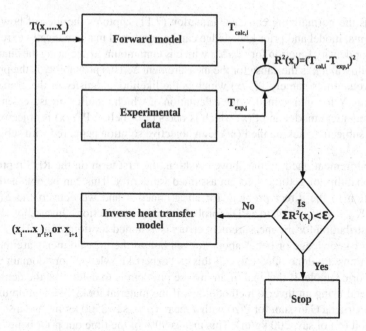

Fig. 7.3 Typical flow chart for solving an inverse problem in heat transfer

posed and suffers from a lack of uniqueness. Several techniques and methods have been developed to address the high dimensionality in these problems and these are elaborated in several books and research articles. However, as engineers, some information on the parameters should be and often is available. The next logical question would be "Why not then make use of it (the available information)?". If the answer to this question is "yes", then we are entering into the territory of Bayesian inference, which is elucidated in the ensuing section.

7.3.1 Bayesian Inference

Bayesian inference is based on the Bayes' conditional probability theorem and uses probability to characterize all forms of uncertainty in a problem. The Baye's theorem relates the experimental data Y(in the previous problem this was T(x)) and the parameters (this was ε in the previous example) and is given by

$$P(x/Y) = \frac{P(Y/x)\ P(x)}{P(Y)} = \frac{P(Y/x)P(x)}{\int P(Y/x)P(x)\mathrm{d}x} \qquad (7.11)$$

In Eq. (7.11), $P(x/Y)$ is called the posterior probability density function (PPDF), $P(Y/x)$ is the likelihood density function, $P(x)$ is the prior density function and

$P(Y)$ is the normalizing constant. Equation (7.11) captures the idea of how measurements, model and prior knowledge can be combined mathematically to return a posterior density function for P(x/Y), which is tantamount to declaring the chance or probability that x is the cause for the measurement Y. This probability is the product of two quantities namely P(Y/x) which is the likelihood density or the probability of getting Y for an assumed x, the calculation of which requires measurements and a mathematical model and P(x) which is one's prior belief. P(Y/x) is objective and P(x) is subjective making the P(x/Y), an objective solution peppered with subjective beliefs.

As aforementioned, in the above equation, the first term on the RHS represents the probability of getting Y for an assumed value of x. This can be obtained from a solution to the direct problem for an assumed x and we convert the $S(X) = \sum_{i=1}^{N}(Y_{\exp,i} - Y_{\sim,i})^2$ in to a PDF (probability density function). Invariably, a Gaussian distribution for the measurement errors is assumed for doing this. The $P(x)$ is our prior knowledge or belief about x, even before the measurements are made or calculations are done. One can call this as "expert knowledge" or "domain expertise". For example, if the goal of an inverse problem is to determine the density(ρ) of a metal using an inverse methodology, if the material looks like aluminium, we can construct a Gaussian for P(ρ) with a mean (μ_p), say 2500 kg/m^3 and a standard deviation (σ) of say, 200 kg/m^3. This means 99% of the time our prior belief is that the density lies between 2500 $\pm$ 600 kg/m^3. This is a fairly reasonable assumption and can help reduce the ill-posedness. Here, if an informative and objective prior like the one mentioned above is used, then the search can be restricted to $1900 \leq \rho \leq 3100$ kg/m^3 instead of $0 \leq \rho \leq \infty$. This is the cornerstone of Bayesian inference wherein engineering knowledge is used to reduce ill-posedness (Balaji 2019).

7.3.2 Steps Involved in Solving a Problem Using Bayesian Approach

The Bayesian method to solve an inverse problem involves three steps:

1. Collection of experimental or measurement data. In the previous example, the data is in the form of temperatures
2. Modelling of (i) the likelihood function that takes into account both the forward model and the measurements and (ii) prior information invariably in the form of a distribution about the parameters to be estimated even before the forward model is solved or the measurements are done. Readers may recall the example of the diagonosis of chest pain by a physician presented earlier.
3. Estimation of x.

The first step is to conduct the experiments and obtain the measured temperatures. In so far as the likelihood is concerned, we exploit the idea of measurement error in temperature as follows:

$$T_{\text{measured}} = T_{\text{simulated}} + \delta \qquad (7.12)$$

In Eq. (7.12), δ is a random variable from a normal distribution with mean "0" and standard deviation σ, where σ is the standard deviation of the measuring instrument (thermocouple). The uncertainty δ is usually assumed to follow a normal or Gaussian distribution, upon which the likelihood can be modelled as

$$P(T/x) = \frac{1}{\left(\sqrt{2\pi\sigma^2}\right)^n} \exp\left(\frac{(T-F(x))^T (T-F(x))}{2\sigma^2}\right) \tag{7.13}$$

In Eq. (7.13) T is a vector of dimension n, i.e. n measurements are available and F(x) is the solution to the forward model with the parameter vector x (x represents a set of parameters). Equation (7.13) can be written as

$$P(T/x) = \frac{1}{\left(\sqrt{2\pi\sigma^2}\right)^n} \exp\left(\frac{-\xi^2}{2}\right) \tag{7.14}$$

$$\text{where } \xi^2 = \sum_{j=1}^{m} \frac{(T_{\text{meas},j} - T_{\text{sim},j})^2}{\sigma^2} \tag{7.15}$$

In Eq. (7.15), $T_{\text{sim},j}$ are the simulated values of T for an assumed X (set of parameters). The posterior probability density function (PPDF) is then given by

$$P(x/T) = \frac{\left[\frac{1}{\left(\sqrt{2\pi\sigma^2}\right)^n} \exp\left(\frac{-\xi^2}{2}\right)\right][P(x)]}{\int \left[\frac{1}{\left(\sqrt{2\pi\sigma^2}\right)^n} \exp\left(\frac{-\xi^2}{2}\right)\right][P(x)]dx} \tag{7.16}$$

In Eq. (7.16), the prior probability density P(x) is usually a standard distribution like a uniform, normal or log-normal distribution. In the case of a uniform prior, P(x) is assigned value (say 1) for all values of x. Such a prior is frequently referred to as non-informative objective prior and will not help us obtain a much sharper PPDF. A sharper PPDF also known as a tighter PDF results in lower standard deviation of the estimates.

Consider the case of P(x) following a normal distribution with mean μ_p and standard deviation σ_p. Mathematically, P(x) is given by

$$P(x) = \frac{1}{\left(\sqrt{2\pi\sigma_p^2}\right)^n} \exp\frac{-(x-\mu_p)^2}{2\sigma_p^2} \tag{7.17}$$

Hence, the PPDF turns out to be

$$P(x/T) = \cfrac{\cfrac{1}{(2\pi)^{\frac{n+1}{2}}(\sigma^n \sigma_p)} \exp(-)\left[\frac{\xi^2}{2} + \left(\frac{(x-\mu_p)^2}{2\sigma_p^2}\right)\right]}{\int \left[\cfrac{1}{(2\pi)^{\frac{n+1}{2}}(\sigma^n \sigma_p)} \exp(-)\left[\frac{\xi^2}{2} + \left(\frac{(x-\mu_p)^2}{2\sigma_p^2}\right)\right]\right] dx} \qquad (7.18)$$

From Eq. (7.18), it is clear that, for every assumed value of the data vector $X(x_1, x_2 \ldots x_n)$, P(x/Y) can be worked out. From this posterior density function, two possible estimates can be pulled out (i) Mean estimate also known as expectation or (ii) Maximum a posteriori (MAP)—which is, the value of x for which P(x/Y) is maximum.

A sampling algorithm is used to generate samples of x. In a multiparameter problem, the marginal PDF of every parameter needs to be computed.

Simplifying Eq. (7.18), we have

$$P(x/Y) = \cfrac{\cfrac{1}{\cancel{(2\pi)^{\frac{n+1}{2}}}(\sigma^n \sigma_p)} \exp(-)\left[\frac{\xi^2}{2} + \left(\frac{(x-\mu_p)^2}{2\sigma_p^2}\right)\right]}{\left[\int \cfrac{1}{\cancel{(2\pi)^{\frac{n+1}{2}}}(\sigma^n \sigma_p)} \exp(-)\left[\frac{\xi^2}{2} + \left(\frac{(x-\mu_p)^2}{2\sigma_p^2}\right)\right]\right] dx} \qquad (7.19)$$

Finally

$$P(x/Y) = \cfrac{\exp(-)\left[\frac{\xi^2}{2} + \frac{(x-\mu_p)^2}{2\sigma_p^2}\right]}{\int \left[\exp(-)\left[\frac{\xi^2}{2} + \frac{(x-\mu_p)^2}{2\sigma_p^2}\right]\right] dx} \qquad (7.20)$$

The expectation or the mean of x is given by

$$\bar{x} = \cfrac{\int x \, \exp(-)\left[\frac{\xi^2}{2} + \frac{(x-\mu_p)^2}{2\sigma_p^2}\right] dx}{\int \left[\exp(-)\left[\frac{\xi^2}{2} + \frac{(x-\mu_p)^2}{2\sigma_p^2}\right]\right] dx} \qquad (7.21)$$

The integral is invariably replaced by a summation for discrete values of x and the expectation turns out to be

$$\bar{x} = \cfrac{\sum_i x_i \, \exp(-)\left[\frac{\xi_i^2}{2} + \frac{(x_i-\mu_p)^2}{2\sigma_p^2}\right] \Delta x_i}{\sum_i \left[\exp(-)\left[\frac{\xi_i^2}{2} + \frac{(x_i-\mu_p)^2}{2\sigma_p^2}\right]\right] \Delta x_i} \qquad (7.22)$$

When Δx_i are the same Eq. (7.22) reduces to the following

$$\bar{x} = \frac{\sum_i x_i \, \exp(-)\left[\frac{\xi_i^2}{2} + \frac{(x_i - \mu_p)^2}{2\sigma_p^2}\right]}{\sum_i \left[\exp(-)\left[\frac{\xi_i^2}{2} + \frac{(x_i - \mu_p)^2}{2\sigma_p^2}\right]\right]} \tag{7.23}$$

$$\sigma_x^2 = \frac{\sum_i (x_i - \bar{x})^2 \, \exp(-)\left[\frac{\xi_i^2}{2} + \frac{(x_i - \mu_p)^2}{2\sigma_p^2}\right]}{\sum_i \left[\exp(-)\left[\frac{\xi_i^2}{2} + \frac{(x_i - \mu_p)^2}{2\sigma_p^2}\right]\right]} \tag{7.24}$$

In Eq. (7.24), σ_x is the standard deviation of the estimated parameter. This is the hallmark of the Bayesian method, as we obtain an estimate and its uncertainty directly and the latter is very hard to determine in many other estimation methodologies.

Example 7.2 Consider Example 7.1, wherein, steady state conduction with Dirichlet boundary conditions on left and Robin condition (mixed condition) on the right side wall was prescribed. Using the same data and sample, determine the mean and standard deviation of the estimated value of emissivity (ε) with Bayesian approach for two cases (i) no prior and (ii) with a Gaussian prior for ε with mean $\mu_{\varepsilon,p} = 0.8$ and $\sigma_p = 0.05$. The standard deviation of uncertainty in measured temperature is 1 K.

Solution:

We can use the Bayesian framework presented above, for the no prior case, to compute the posterior densities for various values of ε_i and these are presented in Table 7.3. The PPDF for the case without prior is shown in Fig. 7.4.

$$\bar{\varepsilon} = \frac{0.231}{0.270} = 0.855, \quad \sigma_\varepsilon = \sqrt{\frac{0.00048}{0.270}} = 0.0421 \tag{7.25}$$

The expectation or mean for the no prior case is 0.855 and the standard deviation of the estimate $\sigma_\varepsilon = 0.0421$. This estimate is also known as the maximum likelihood estimate.

Next, we include the Gaussian prior and obtain the results as presented in Table 7.4.

From Table 7.4, the expectation (or mean) of ε and also the standard deviation of the estimate are obtained as

$$\bar{\varepsilon} = \frac{0.1452}{0.1724} = 0.84, \quad \sigma_\varepsilon = \sqrt{\frac{0.000144}{0.1724}} = 0.028 \tag{7.26}$$

The PPDF for this case with the Gaussian prior is given in Fig. 7.5.

Upon ingestion of the Gaussian prior, the standard deviation of the estimate of ε has decreased substantially. The informative and subjective Gaussian prior has thus been extremely useful in the estimation process.

It is possible for us to use the Markov chain, wherein, the next sample of x (ε in this case) depends on only the current value of x. This can be accomplished by drawing the new sample from a Gaussian distribution with its mean being the current

Table 7.3 Estimation of ε using the Bayesian method (no priors)

S.No	ε_i	$S(\varepsilon_i)$	$\varepsilon_i \exp^{-\left(\frac{S(\varepsilon_i)}{2\sigma^2}\right)}$	$\exp^{-\left(\frac{S(\varepsilon_i)}{2\sigma^2}\right)}$	$(\varepsilon_i - \bar{\varepsilon})^2 \exp^{-\left(\frac{S(\varepsilon_i)}{2\sigma^2}\right)}$
1	0.15	386.02	2.24×10^{-85}	1.49×10^{-84}	7.39×10^{-85}
2	0.25	259.05	1.39×10^{-57}	5.58×10^{-57}	2.03×10^{-57}
3	0.35	169.75	3.55×10^{-37}	1.015×10^{-36}	2.57×10^{-37}
4	0.45	98.84	1.55×10^{-22}	3.45×10^{-22}	5.59×10^{-23}
5	0.55	52.84	1.84×10^{-12}	3.35×10^{-12}	3.08×10^{-13}
6	0.65	23.57	4.93×10^{-6}	7.6×10^{-6}	3.13×10^{-7}
7	0.75	7.81	0.015	0.020	0.0002
8	0.85	3.01	0.189	0.222	1.937×10^{-6}
8	0.95	7.14	0.027	0.0281	0.0002
Σ			0.231	0.270	0.00048

Fig. 7.4 PPDF of ε with the Bayesian method (no prior) for example 7.2

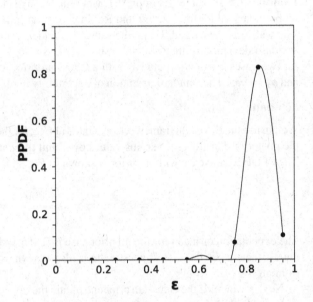

value of "x" and "σ" being typically 5% of the current mean. While a sample with higher PPDF is always accepted, rejection is done with a probability based on an acceptance ratio (see Balaji 2019 for a further discussion on this). This method is known as Metropolis Hastings (MH) based "Markov chain Monte Carlo (MCMC)" method. For further discussions on the powerful MCMC method readers may refer to statistics books and journals.

Table 7.4 Estimation of ε using the Bayesian method (with a Gaussian prior) for example 7.2

ε_i	$S(\varepsilon_i)$	$A=\dfrac{S(\varepsilon_i)}{2\sigma^2}$	$B=\dfrac{(\varepsilon-\mu_{\varepsilon,\text{prior}})^2}{2\sigma_p^2}$	$\varepsilon_i\exp^{-(A+B)}$	$\exp^{-(A+B)}$	$(\varepsilon_i-\bar\varepsilon)^2\exp^{-(A+B)}$
0.15	386.02	193.01	58.68	7.35×10^{-111}	4.901×10^{-110}	2.35×10^{-110}
0.25	259.05	129.52	42.01	7.92×10^{-76}	3.168×10^{-75}	1.50×10^{-75}
0.35	169.75	82.87	28.12	2.17×10^{-49}	6.198×10^{-49}	1.50×10^{-49}
0.45	98.84	49.42	17.0	6.33×10^{-30}	1.41×10^{-29}	2.16×10^{-3}
0.55	52.84	26.42	0	3.14×10^{-16}	5.7×10^{-16}	4.88×10^{-17}
0.65	23.57	11.78	0.5	2.17×10^{-7}	3.34×10^{-7}	1.24×10^{-8}
0.75	7.81	3.91	0.347	0.0106	0.0142	0.00012
0.85	3.01	1.50	0.347	0.1334	0.1569	8.89×10^{-6}
0.95	7.14	3.57	3.125	0.0012	0.0012	1.42×10^{-5}
		Σ		0.1452	0.1724	0.000144

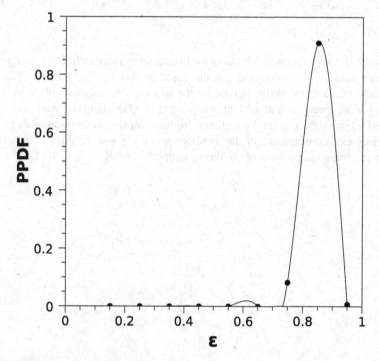

Fig. 7.5 PPDF of ε with the Bayesian method (for a Gaussian prior) for example 7.2.

Problems

7.1 Consider a thin aluminium foil coated with a paint of "high" emissivity ϵ with dimensions of 2 cm × 2 cm and 2 mm thickness suspended in an evacuated chamber. The chamber is maintained at 303 K and the foil is initially at 373 K. The foil gets cooled radiatively and its measured temperature response is tabulated below (Table 7.5). Estimate the emissivity of the coating by using an exhaustive equal interval search, in the range $0.65 \le \varepsilon \le 0.95$ with an interval of 0.05 and then switching to a Lagrangian interpolation formula by using a least square approach. The foil density is 2707 kg/m^3 and the specific heat is 903 J/kgK.

Table 7.5 Temperature time history of the aluminium foil under consideration in Problem 7.1

t, (s)	50	100	150	200	250	300	350
T, (K)	363.3	355.3	348.6	342.9	338.0	333.8	330.2

7.2 Consider the problem of determination of emissivity of a thin foil with a measured temperature distribution that was discussed in exercise Problem 7.1. With the same data and sampling, determine the mean of the estimate of "ε" using a Bayesian approach with (i) Uniform prior (ii)A Gaussian prior with $\mu_p = 0.84$ and $\sigma_p = 0.06$. The total uncertainty in the temperature measurement (which arises as a consequence of the thermocouple error and the error in accurately determining the position of the thermocouple) is ±1 K.

Bibliography

Balaji C., 2019, Thermal system design and optimization, Second Edition, CRC Press, New Delhi.

Howell J. R., R. Siegel, M. P. Menguc, 2011, Thermal radiation heat transfer, Ane Books, Boca Raton.

Incropera F. P., D. P. Dewitt, T. L. Bergman, A. S. Lavine, 2007, Fundamentals of Heat and Mass Transfer, John Wiley, New York.

Leckner, B., 1972. Spectral and total emissivity of water vapor and carbon dioxide. Combustion and flame, 19(1), pp.33-48.

Liou K. N., 2002, An introduction to atmospheric radiation, Academic Press, London.

Mahan J. R., 2002, Radiation heat transfer- A statistical approach, John Wiley, New York.

Modest M. F., 2003, Radiative heat transfer, Academic Press, London.

Muralidhar K., J. Banerjee, 2010, Conduction and Radiation, Narosa Publishing House, New Delhi.

Sparrow, E. M., J. L. Gregg, 1956, Laminar free convection from a vertical plate with uniform surface heat flux, Transactions of the ASME, 78, 435-440.

Venkateshan S. P., 2009, Heat Transfer, Ane Books, New Delhi.

Wallace J. M., P. V. Hobbs, 2006, Atmospheric Science- An introductory survey, Academic Press, London.

© The Author(s) 2021
C. Balaji, *Essentials of Radiation Heat Transfer*,
https://doi.org/10.1007/978-3-030-62617-4

Index

A
Absorptivity, 72

B
Bayesian inference, 201
Bayes' theorem, 201
Beer's law, 158
Blackbody, 57
Boltzmann constant, 196

C
Conduction, 1, 10, 14, 19, 24, 67, 71, 74, 78
Contour integration, 117
Convection, 1–5, 10, 14, 19, 23, 24, 65, 67, 71, 74, 85, 86, 96, 97

D
Decomposition rule, 119
Diffuse surface, 57, 59, 60, 62, 64, 65, 72, 76, 77, 81
Directional total emissivity, 62

E
Effective emissivity, 139
Electromagnetic spectrum, 6, 10, 45, 186
Emissivity, 2, 3
Enclosure theory, 99
Exponential integral, 162

F
Far infrared radiation, 185
First radiation constant, 44

F
Frequency, 7

G
Gas mixture, 181
Gaussian distribution, 202
Gaussian prior, 205

H
Harmonic oscillator, 33, 38
Hemispherical spectral emissivity, 61
Hemispherical total absorptivity, 80
Hemispherical total emissivity, 63, 65, 66, 69
Hohlraum, 12
Hottel's crossed string method, 111

I
Irradiation, 129, 130

L
Lambert's law, 158
Law of corresponding corners, 121
Line by line model, 176
Log-normal, 203

M
Maximum a posteriori, 204
Maxwell–Boltzmann statistics, 38
Mean beam length, 164
Microwave radiation, 185
Multi spectral instrument, 153

© The Author(s) 2021
C. Balaji, *Essentials of Radiation Heat Transfer*,
https://doi.org/10.1007/978-3-030-62617-4

Printed in the United States
by Baker & Taylor Publisher Services